MARCH 1974
VOLUME XXVII, No. 3   ISSUE No. 107

the review of metaphysics

a philosophical quarterly

A COMMEMORATIVE ISSUE

THOMAS AQUINAS

1224-1274

articles

Cornelio Fabro, *The Intensive Hermeneutics of Thomistic Philosophy: The Notion of Participation* 449

Kenneth L. Schmitz, *Enriching the Copula* 492

Matthew R. Cosgrove, *Thomas Aquinas on Anselm's Argument* 513

Edward P. Mahoney, *Saint Thomas and Siger of Brabant Revisited* . . 531

Fernand Van Steenberghen, *The Problem of the Existence of God in Saint Thomas' COMMENTARY ON THE METAPHYSICS of Aristotle* 554

William A. Wallace, *Aquinas on the Temporal Relation Between Cause and Effect* 569

John F. Wippel, *The Title FIRST PHILOSOPHY according to Thomas Aquinas and his Different Justifications for the Same* . . 585

books received

William A. Frank and Staff, *Summaries and Comments* 601

philosophical abstracts 628

announcements 651

ARTICLES

THE INTENSIVE HERMENEUTICS OF THOMISTIC PHILOSOPHY
The Notion of Participation

CORNELIO FABRO

Translated by B. M. BONANSEA

I

IN THE PLATONIC TRADITION, the term "participation" signifies the fundamental relationship of both structure and dependence in the dialectic of the many in relation to the One and of the different in relation to the Identical, whereas in Christian philosophy it signifies the total dependence of the creature on its Creator. The term participation has played an extensive role in Patristic and medieval speculation.

The starting point or stimulus for an interpretation of Thomism by means of the notion of participation—an interpretation that was unusual and entirely neglected by the tradition—was first of all the controversy about the so-called "analytic nature" of the principle of causality. This arose toward the end of the last century and continued through the first decades of our century within Neoscholastic (and Neothomistic) philosophy. Although some philosophers (De Margerie, Fuzier, Laminne) had already defended the synthetic nature of the principle of causality, that is, its novelty and consequent irreducibility to the principle of contradiction, nevertheless the rationalistic solution was always dominant. This rationalistic tradition conceived the principle of causality as derived from the principle of sufficient reason, which in turn was derived from the principle of identity-contradiction (Descoqs, Maritain, Garrigou-Lagrange). On the other hand, the "synthetists" (e.g., Sawicki, Hessen, and Geyser) placed themselves directly behind Kant and Schopenhauer and openly broke with the traditional interpretation which they accused of subordinating the understanding of reality to abstraction.[1] In so doing, the synthetists made no secret of their adherence to the Kantian apriori. A new way had therefore to be sought to resolve the *impasse*.

Yet the convincing and decisive moment was the realization that, in an attempt to solve crucial issues of the constitutive relation between God and creatures, between the Infinite and the finite—such as those concerning total dependence (creation and divine motion), radical structure (composition of *essentia* and *esse*) and fundamental semantics (analogy)—St. Thomas had placed the Platonic notion of participation at the very foundation of the Aristotelian couplet of act and potency. The theory was then advanced, which both text and context have supported and clarified, that the very notion of *ens* and that of *esse* as intensive emerging act, sprang in Thomism from within that same notion of participation and marked the definitive overthrow of both classic and scholastic essentialism (formalism). The real composition of *essentia* and *esse*, fought against by scholastic philosophy and diverted by historical Thomism into the modal distinction of *essentia-existentia*, was thus placed at the center of the Thomistic interpretation of the real [2] on the level of the transcendentality of Hegel's *Diremtion* and Heidegger's *ontologische Differenz*.

Extensive and prolonged research, spurred by the insistence upon interiority in modern throught, led to the discovery of a corresponding perfect coherence in the structure of causality.[3] The newly

---

[1] The first research, which resulted in the discovery of the formula of participation, was stimulated by the theme of a contest announced by the Pontifical Academy of St. Thomas in 1934: "The principle of causality: its psychological origin, its philosophical formulation, and its necessary and universal value." The content of my study, awarded first prize by the Academy, was published under the title, "La difesa critica del principio di causa," in *Rivista di filosofia neoscolastica*, XXVIII (1936), 102–41 and reprinted in my volume, *Esegesi tomistica* (Rome, 1969), pp. 1–48.

[2] The constitutive participation of the structure of the real is the subject of my work, first published in 1939, *La nozione metafisica di partecipazione secondo S. Tommaso d'Aquino* (3d ed.; Turin, 1963), which includes three distinct parts: I. Historical foundation of the notion of participation; II. Fundamental implications of the Thomistic notion of participation; III. Inner development and contents of the Thomistic notion of participation.

[3] The complex and difficult notion of dynamic participation is the subject matter of my second volume devoted to the study of Thomistic participation, *Participation et causalité* (Paris-Louvain, 1960; Italian edition, Turin, 1961), which contains the results of twenty years of research. The

advanced theory found its confirmation in the continuous and ever-increasing references in Thomistic texts to the Neoplatonic tradition, especially that deriving from Proclus (Pseudo Dionysius, *De Causis*). Yet this latter tradition was itself overturned by the elevation of the *ens-esse* over and above the Good and by the consequent emerging synthesis of *essentia-esse* and of the formal-exemplary causality of Plato with the formal-efficient causality of Aristotle. This was therefore a new synthesis [4] and such that it could not be reduced either to Plato or Aristotle, not to Hellenistic or Christian Neoplatonism.

It is true that the importance of Aquinas' thought in the evolution of the human spirit is beyond question, just as is the importance of St. Augustine's thought. However, it must be admitted that neither the Augustinian nor the Thomistic school has always resisted the desire for system (*concupiscentia systematis*), and for the best known Thomists this is a system with an ever more noticeable Aristotelian background. The endless discussions from the 14th to the 16th century between Thomists [5] and anti-Thomists offer an impressive array of arguments aiming at the defense of the theses of their respective schools, rather than the dispassionate study of the much more important issue of the "foundation" and original meaning of Aquinas' attempt at a *new* synthesis: the speculative meeting of Plato and Aristotle on the level of Christian creationism. It is in this sense that the new *Thomas-Forschung* has resolutely moved in

---

work includes the following parts: I. Formation of Thomistic *esse*; II. Causality of *esse*; III. Dialectic of causality.

[4] In his work, *Die neuplatonische Seinsphilosophie und ihre Wirkung auf Thomas von Aquin* (Leiden, 1966), the distinguished scholar K. Kremer attempts to reduce Thomism to Dionysian Neoplatonism along the lines of the School of Chartres, with a view to achieving a synthesis of Thomism and immanentism. In the author's view, Christian Neoplatonism would maintain that God himself is the *Esse* of creatures according to the classic formula of Pseudo Dionysius: Αὐτός 'εστι τὸ εἶναι τοῖς οὖσι (*De divinis nominibus*, c. V, par. 4; PG 3, 817 D). Kremer admits therefore that his interpretation must leave aside (!) the Thomistic distinction of *essentia* and *esse*. See our critique in "Platonism, Neo-Platonism, Thomism," *The New Scholasticism*, XLIV, No. 1 (1970), pp. 69–100. The article appeared also in my volume, *Tomismo e pensiero moderno* (Rome, 1969), pp. 435–60.

[5] For the history and nature of these controversies see C. Fabro, "L'obscurcissement de l'*esse* dans l'école thomiste," *Revue Thomiste*, LXIII, No. 3 (1958), pp. 443–72. See also Fabro, *Participation et causalité*, pp. 280ff.; It. ed., pp. 603ff.

our direction:

> On the other hand, the doctrine of participation, which is clearly derived from Neoplatonism, has scarcely been mentioned by those of his [Thomas'] interpreters who have insisted on its Aristotelian origin. This same doctrine is considered as central and analyzed in great detail by many recent commentators, who are not afraid to see in it, along with Aristotelian and Christian features, important elements of Neoplatonic origin. This is not surprising, since many texts of Proclus and the *Liber de Causis*, in addition to writings by St. Augustine, Boethius, the Areopagite, and many Arabic sources following the same line of thought, were available to him.[6]

But the journey in this speculatively radical direction has just begun; besides, it must overcome the resistance of a prevailing formalistic tradition.

One philosopher who was decidedly on his way to seeing in Thomas the metaphysical tension of act and potency grounded in the notion of participation, was Marsilio Ficino, the prince of Humanism of the *Quattrocento*. In his major work, the *Theologia Platonica*,[7] he acknowledges the principal theses of Thomistic metaphysics and, while recalling the notion of participation, he fights against the triumphant Averroism with the same arguments used by St. Thomas, whom he calls "the bright light of Christian theology." Had his voice been heard, many useless disputes within the orders would have been avoided.

---

[6] See P. O. Kristeller, *Le Thomisme et la pensée italienne de la Renaissance* (Montréal-Paris, 1967), pp. 24, 29. In n. 12 of p. 29, which lists the bibliography on Thomistic participation, the author characterizes the study of R. J. Henle, *Saint Thomas and Platonism* (The Hague, 1956), as *"fort utile"* but such that *"se limite à une étude des textes de saint Thomas où il parle explicitement de Platon ou des platoniciens."* The author shows an even greater disappointment with Gilson's position on this question: *"Gilson traite de la participation chez saint Thomas seulement dans une petite note (Le Thomisme* [Paris, 1948], p. 182, n. 3) *et affirme 'que participer, en langage thomiste, ne signifie pas être une chose, mais ne pas l'être'. Or, en langage platonicien, participer signifie et être une chose et ne pas l'être en même temps, et la ressemblance d'une copie à son archétype ne comporte pas une ressemblance réciproque."* It will be seen presently that "to participate," in a Thomistic sense, is but a continuous growing of positivity on all levels of being and acting.

[7] P. O. Kristeller must again be congratulated for having thrown some light on this point. See *Il pensiero filosofico di Marsilio Ficino* (Florence, 1953), pp. 147, 159ff., 184, 192, 208 and *passim*. For a direct comparison between the doctrines and texts of Aquinas and M. Ficino see C. Fabro, "Influenze tomistiche nella filosofia del Ficino," *Studia Patavina*, III (1959), pp. 396–413. Reprinted in *Esegesi tomistica*, pp. 313–28.

II

In ordinary usage participation has a sociological meaning that can be extended to include all relations pertaining to the union of the members of a collectivity for the objectives and purposes that concern them in one way or another. This kind of participation may refer to knowledge or will, sentiment or activity, as the following expressions indicate: to participate in news, a happy or unhappy event. Such applies not to any person whatsoever but to one who can or must "take part," that is, associate himself at least with the feelings of him who communicates the news or the event. Participation may also mean collaboration or joining together in the performance of a task for the achievement of a common objective. In this sense one is said to participate in an action, good or bad, in an undertaking, or in an enterprise. This is a kind of physical and moral solidarity that involves participation in the responsibility for the results of the collaboration itself. We likewise say, phenomenlogically speaking, that one participates in the joy or sadness of someone else, in his projects or convictions, to express a subject's association with the particular situation of another subject in the most intimate nucleus of his personality. Thus participation is the broadest and most effective term to express the relation of consciousness to reality in its manifold aspects: physical, moral, cognitive, artistic, sociological, economic, and so forth.

When considered from the semantic point of view, participation can be taken in an active transitive sense and then it means to give and communicate something to someone else, or to share something with someone else. In this sense we also say that God participates his being, his goodness, his truth to creatures. Taken in its intransitive sense, participation means to share or take part in something. It can be said then that participation embraces all kinds of objects from the physical world to moral life, from extended quantity to the life of the spirit. The only difference is that, whereas in the quantitative and material order participation attains directly to the object inasmuch as a certain "whole" is being divided and distributed in its parts, in the moral and in the strictly metaphysical order participation concerns properly speaking the *mode* of having and receiving, in the sense that the "whole" remains intact and undivided, while an aspect or form of the object is being participated.

As far as the Latin term is concerned, participation means to take part (*partem capere*) or to have part (*partem habere*—German: *teilnehmen, teilhaben*). "To participate is like taking a part; thus when something receives a part of what belongs to another fully, it is said to participate in it." [8] Grammatically speaking, we have the so-called participle, which acts like a mediator between the noun and the verb: "A participle is that which partakes of the noun and the verb. It partakes of the noun with regard to genera and cases, of the verb with regard to tenses and meanings, and of both with regard to number and figure." [9]

The notion of participation was introduced into philosophy by Plato to explain the relation of the concrete sensible to the separate Intelligible (the Idea), of the singular to the Universal. However, Aristotle ascribes the first origin of the doctrine of participation to the Pythagoreans, or more generally, to the converging contributions resulting from the conflict between the solutions of Parmenides, Heraclitus, Pythagoras, and their followers, which served as a springboard for Socrates' philosophy.[10] In fact, in his search for the foundation of virtue, Socrates concluded that it had to be immutable, one, simple, and therefore have the characteristics of a Universal that exists in itself (τὸ καθόλου) and apart from the sensible objects which are constantly changing. Plato accepted the Socratic solution and called the Universal "Idea" (ἰδέα, εἶδος). He placed all sensible things outside the Idea and explained the relation of sensible things (the "many") to the Idea (the One) by the doctrine of participation. This means that all things that have the same name share the common nature of the species. As for Plato's contribution to the notion of participation, Aristotle asserts that he only changed the name (τοὔνομα μετέβαλεν). Whereas the Pythagoreans said that things are or exist by imitation (μίμησις) of numbers, Plato coined a new term and said that they are or exist by participation (μέθεξις). Yet neither side, in Aristotle's view, cared to explain what imitation or participation really meant.[11]

---

[8] St. Thomas, *In I Boeth. de Hebd.*, 1. 2, n. 24; ed. Taur., p. 396 *b*.
[9] S. Isidorus Hisp., *Etymol.* I, 21, c. II; PL 82, col. 88.
[10] Aristotle, *Metaph.* I, 5, 986 *b* 18ff. An accurate exposition of the sources of the notion of participation has been made by Sister M. Annice in her article, "Historical Sketch of the Theory of Participation," *The New Scholasticism*, XXVI (1952), pp. 49–79.
[11] Aristotle, *Metaph.* I, 6, 987 *b* 10ff.

The truth of the matter is that participation is found in Plato only in its nascent state, and receives a whole gamut of modes of having and being according to a certain relationship of dependence, similarity, coexistence, and the like. Already in the *Phaedo* things are said to participate when they "acquire" [12] or "receive" [13] something, or simply when they are what they are by a "presence" or "communion" [14] or even by "appertaining," [15] with respect to the "model" in which many participate.[16] Thus the subjects of participation are the many and the manifold.[17]

The notion of participation was meant to be the solution to Zeno's problem: if many things exist, they must simultaneously be similar and dissimilar, one and many, at rest and in motion.[18] The answer to this problem is that the "Ideas" do not mix with sensible things but exist "by themselves" (καθ'αὐτά) and are "separate." [19] According to Aristotle, who must have been aware of the oral teaching of his master, Plato had inserted the mathematical beings [20] as intermediaries (τὰ μεταξύ) between the sensible objects and the separate forms. By this metaphysical mathematization attributed to him by Aristotle, Plato remained faithful to his fundamental inspiration of transforming mathematical relations into relations of constitutive logic. Indeed for Plato the object of knowledge is either the idea as an "exemplar" (παράδειγμα) or "that which completely is" (τὸ παντελῶς ὄν) and is therefore perfectly knowable (τὸ παντελῶς γνωστόν), so that it is the "one among the many" (τὸ ἓν ἐπὶ τῶν πολλῶν) and makes possible the knowledge from above of what is transitory and corruptible.[21] To the principal difficulties raised against participation in its twofold relationship of the intelligible to the sensibles and of the intelligibles (the genera) among themselves, Plato gave in his later dialogue a twofold answer of decisive importance for

---

[12] μετάσχεσις: *Phaedo* 100 c, 101 c.
[13] μετάληψις: 102 b.
[14] παρουσία, κοινωνία: 103 d.
[15] ἐνεῖναι, παραγίγνεσθαι: 103 d, 105 c. Cf. *Sympos.* 211 b.
[16] τὰ τοῦ αὐτοῦ μετέχοντα τύπου: *Rep.* 402 d.
[17] εἶναί τι ἕκαστον τῶν εἰδῶν καὶ τούτων τἆλλα μεταλαμβάνοντα πολλά . . . : *Phaedo* 78 d. Cf. *Rep.* 476 a and ff., 479 a, 507 b.
[18] *Parm.* 127ff.
[19] χωρίς: *Phaedo* 129 de.
[20] *Metaph.* I, 6, 987 b 12ff.
[21] τὸ νοεῖν τι φθαρέντος: *Soph.* 248 e.

the development of philosophy, including Aristotelian philosophy. On the one hand, he admitted the existence of non-being (μὴ ὄν) as the "subject" of participation or imitation, and on the other hand, he extended the relation of participation within the very realm of the Ideas (κοινωνία τῶν γενῶν) so as to make possible a multiple participation.[22]

One of the major difficulties inherent in the Platonic notion of participation based on the logico-mathematical relation of the universal to the particular as imitation-similitude, is the celebrated "third man" argument (τρίτος ἄνθρωπος), which is mentioned by Aristotle [23] but had already been discussed by Plato himself.[24] It consists in this. If the similarity among many sensibles presupposes the "form in itself," then the similarity of the many to the forms presupposes another form, and so forth.[25] To this objection Plato has no answer. Aristotle's reaction to the Platonic notion of participation is found in the works of his maturity, where he shows great firmness and an attitude that comes close to contempt.[26] However, one must bear in mind that Aristotle came to realize the inadequacy of the Platonic doctrine of participation only gradually, since he himself had adhered to it in his youthful dialogues of Platonic inspiration. The very titles of these dialogues (*Eudemus, Symposion, Eroticus, Protrepticus, Politicus, Sophistes,* etc.) [27] seem to bear out this view. Moreover, traces of that doctrine can still be found in the *Organon* where he discusses the logical relation of universals, i.e., of individuals to species and of species to genus.[28] From these latter there arose that important nucleus of the Thomistic doctrine of predicamental participation.

---

[22] μετέχειν δὲ πολλῶν οὐδὲν χωλύει: *Parm.* 161 a. Cf. P. Natorp, *Platos Ideenlehre* (Leipzig, 1921), pp. 231ff., 469f.

[23] *Metaph.* I, 9, 990 b 17 and XII, 1, 1059 b 8.

[24] *Parm.* 132 ab.

[25] See the ample exposition of the argument by Alexander of Aphrodisias, *In Metaph.* 990 a 15; Hayduck 83, 34–84, 1ff.

[26] μεταφορὰς λέγειν ποιητικάς: *Metaph.* I, 9, 991 a 20. See also pertinent texts in *Metaph.* XIII, 4, 1079 a 4ff., 5, 1079 b 24ff.

[27] *De Philosophia.* See *Aristotelis Dialogorum Fragmenta,* ed. R. Walzer (Florence, 1934).

[28] *Topic.* IV, 1, 121 a 11ff. See Fabro, *La nozione metafisica di partecipazione,* pp. 145ff.

III

To the Platonic doctrine of participation based on imitation and transcendence, Aristotle opposed the *immanence* of the form in sensible substances and the *causality* of the individual singular in the process of natural becoming. At this point Greek thought did not seem to offer any way out of the *impasse* in which it found itself. Indeed, if Aristotle was right in vindicating the reality of sensible substance, this latter could not be viewed by him as the perfect substance. As a matter of fact, he himself admitted above the material world the existence of spiritual substances, whether Intelligences or souls, as moving principles of the stars.[29] Nevertheless the radical opposition between Platonic participation and Aristotelian causality rendered impossible an adequate explanation of the foundation of the real. From a dialectical standpoint, a solution could be reached either by meeting the requirements of causality from within the doctrine of participation, or by clarifying participation by means of causality. This latter has been the course followed by pure Aristotelianism (e.g., Alexander of Aphrodisias), whereas the former has been the way followed by Neoplatonism and the one that has exerted a greater influence on the shaping of Western Christian thought. This derivation is expressly affirmed by Eusebius, who, despite his usual critical attitude toward Aristotle, dwells on the subject with a certain delight.[30]

The endeavor as well as the distinctive feature of Neoplatonism is to show the fundamental agreement or harmony between Plato and Aristotle and to explain Aristotle's anti-Platonic polemics solely in terms of Plato's inadequate mode of expression (through poetic metaphors, myths, and the like) rather than in terms of doctrinal differences in the solution of fundamental problems. In the struggle between Neoplatonism and the fast-growing Christianity, the attempt to establish major agreement between the two greatest Greek philosophers that would go beyond their verbal differences, amounted to no less than a defense of the Greek ideal of wisdom against a foreign religion based on purification from sin by means of Christ's redemption.

---

[29] See *De caelo*, III, 2, 285 a 27ff.
[30] *Praep. Ev.* XV, 7; CB 43, 2, Mras II, 365, 5ff.

It can be said that Neoplatonism, while adhering to the fundamental principle of participation, transforms its structure by absorbing from within the Aristotelian criticism together with its underlying principles. The harmony between Plato and Aristotle is already asserted by Ammonius, Plotinus' teacher, who, in a fragment of a dialogue on Providence written by Hierocles, is credited with the explicit use of the "intensive method." This consists in a higher synthesis whereby the fundamental principles of two opposing doctrines are reconciled.[31] This new criterion is explicitly assumed in the Latin world by Boethius, who transmits it to the Christian Middle Ages both as a task and as a program. He writes:

> I will translate into the Roman language all Aristotle's works that will fall into my hands, as well as their commentaries. . . . I will also translate into Latin and comment upon all the dialogues of Plato. Once this is done, I will not disdain to bring the statements of Aristotle and Plato into some sort of a harmony, in an attempt to show that, far from disagreeing in everything, they agree in many things, and especially in their philosophical teaching.[32]

The theme of a doctrinal agreement in the teaching of Plato and Aristotle is handed down to Arabic Neoplatonism almost systematically and is extended to all sections of philosophy, to the point of reducing the difference between the two philosophers, as Alfarabi indicates, to one of method. Whereas Plato has chosen analysis, Aristotle has opted for synthesis. Thus Aristotle becomes "the follower and perfecter, the helper and counselor of Plato."[33] In the Renaissance this conciliatory tradition is being constantly upheld (Gennadius or George Scholarios, Bessarion) and a campaign is waged against those who, following the lead of Alexander (Plethon), cling to the theory of opposition and assert the incompatibility of the teachings of the two philosophers.[34]

The endeavor to work out an agreement between Platonic transcendence and Aristotelian immanence has a decisive effect on the metaphysical notion of participation. It involves certain funda-

---

[31] See Photius, Bibl., French trans. by M. N. Bouillet, *Les Ennéades de Plotin* (Paris, 1857), I, pp. xcivf.
[32] Boeth., *In 1. Arist. De interpretatione*, 1.II, prol.; PL 64, 433.
[33] F. Dieterici, *Alfarabi's philosophische Abhandlungen* (Leiden, 1892), pp. 3, 17ff.
[34] See B. Tatakis, *La Philosophie byzantine* (Paris, 1949), pp. 281ff. The study of P. O. Kristeller, "Byzantine and Western Platonism in the Fifteenth Century," in *Renaissance Concepts of Man* (New York, 1972), pp. 86–109, is more precise and better documented.

mental steps according to which the "convergences" between the two philosophers are brought out and set in relation to ever more comprehensive and unifying principles as transition is made from Greek Neoplatonism to rigid Islamic monotheism and ultimately to the creationist theory of Christianity. Such processes and attempts remain, however, at a purely experimental stage and assume the form of a more or less eclectic reconciliation until Aquinas introduces his original notion of *esse* as an intensive act which offers the ultimate ground for the doctrine of participation.

Indeed, by the celebrated doctrine of the three Hypostases of Intellect, Soul, and Life (νοῦς, ψυχή, Ζωή), an attempt is made by Plotinus and Neoplatonism in general to reduce to a minimum the distance between transcendence and immanence. Thus Aristotle's νοῦς is made to coincide, as far as its contents is concerned, with Plato's ὄντως ὄν, inasmuch as it itself is a multiplicity of ideas. Just as without the Ideas there is no true science or true reality, so without Intellect (Aristotle's νοῦς) there can be neither science nor reality. Intellect is, in effect, the supreme principle of the world inasmuch as, by thinking itself, it thinks of a "multiplicity." Without the multiple, thinking is indeed impossible. In turn, this multiplicity of "thoughts" cannot be derived by Intellect from the sensible world but solely from itself. The existence of Plato's Ideas (the εἴδη νοητά) is thus justified through a process whereby Aristotle is transformed, or rather overturned, by means of an Aristotelian principle.

Once the obstacle of the "separation" (χωρισμός) of the Ideas from things is removed, Plotinus faces the difficulty offered by the lack of causality in the Ideas. He decides to resolve this through his doctrine of emanation (πρόοδος) and particularly through his doctrine of "World soul." He teaches that the Ideas, which proceed from νοῦς, are given, as it were, to the *matter* of the soul, and that the soul, as a principle of movement, is also responsible for the origin and destruction of the things formed by the Ideas in matter. On the other hand, the Ideas themselves, according to Plotinus, are endowed with quality and quantity, motion and rest, so that sensible things depend on the Ideas, from which they derive their movements and changes of quality and quantity according to their nature.[35]

---

[35] See A. Covotti, *Da Aristotele ai Bizantini* (Naples, 1935), pp. 226ff.

Yet the syncretism of the Neoplatonic notion of participation has a more profound meaning than may at first appear. It aims to go beyond Plato and Aristotle and regain the fundamental truths of Presocratic philosophy represented by men such as Empedocles and Pythagoras, and especially by Parmenides, from whom the problematic of the notion of participation took its starting point. It is within this trend of thought that the "dialectical method" attains prominence and reaches its highest point in Proclus, whose triadic process starts from the One, goes on to multiplicity, and ends up in Being. The novelty of this dialectic, explicitly introduced into the notion of participation, consists in the importance given to "negativity" as a moment within the process and consequently as the foundation of the dialectic itself.[36] The negations (ἀποφάσεις) are in effect not to be taken merely as "privations" but rather as productions that determine their contraries, in accord with the principles laid down in Plato's *Parmenides*. Proclus affirms indeed that "the method of negations (τρόπος τῶν 'αποφάσεων) is excellent; it conforms to the dignity of the One; its function is primary; it far transcends all things in the unknowable and ineffable excellence of simplicity."[37] Proclus' work represents the greatest attempt at a solution of the speculative antithesis between Plato and Aristotle and replaces the negative alternative proposed by Alexander.[38]

In general it can be said that Arabic Neoplatonism develops the notion of participation mainly through the ascending (toward the One) intuitive method that stems from Plotinus and Porphyry and is handed down through Augustine to the medieval schools directly inspired by him. The Thomistic notion of participation, on the other hand, takes its direct inspiration from the more rigorous descending dialectical method of Jamblichus and Proclus through the intermediation of the speculation of Pseudo-Dionysius and the Arabic opuscule *De Caus s*.

In their attempt to harmonize the contrasting viewpoints, the Neoplatonists had been preceded by the Stoic Poseidonius, who,

---

[36] See Hegel, *Geschichte der Philosophie*, ed. Michelet (Berlin, 1844), II, p. 66.

[37] *Theol. Plat.*, 1.II, c. 10; ed. Portus (Hamburg, 1618), fol. 109.

[38] See the explicit statement of Simplicius, *In de caelo*, III, 7, 306 *a* 1; ed. J. J. Heiberg, 640, 20ff. The text is reported in full in my volume, *La nozione metafisica di partecipazione*, p. 60, n. 1.

especially in his commentary on the *Timaeus*, had already endeavored to achieve a synthesis between theology and mathematics.[39] The speculative innovation of the new approach to the notion of participation worked out by Neoplatonism in its two main directions has had a more or less noticeable influence on modern thought, as will be seen in the course of this paper. Moreover, the originality of, as well as need for, such an innovation is far from having been exhausted. It consists, first, in the fact that it does away with the fracture between Being and the One, and secondly, in the realization that knowledge has been raised to the level of being by developing its own unlimited power (*vis*) within the plenitude of being and life. Thus Neoplatonism has succeeded in bringing about a theoretical development, completed in its own way, of the Parmenidean principle of the relation between being and thought. It is worth noting in this connection that through the new notion of participation brought about by Neoplatonism, the mediation of the contrast between Plato and Aristotle is effected in its crucial point (the doctrine of νοῦς) by direct recourse to Aristotle himself. That is why more recent historiography, especially under the influence (positive and negative) of W. Jaeger, accepts the theory of continuous development from Plato to Aristotle and Neoplatonism, in such a way that Aristotelian metaphysics itself appears to be profoundly indebted to Platonism. "What has for centuries been interpreted as general metaphysics ('doctrine of being-as-such') originated in Aristotle as another presentation of this [Platonic] excessively realistic *Ableitungssystem*."[40]

IV

St. Augustine developed his philosophy outside the Proclian line of thought mentioned above. Not knowing the works of Aristotle, he could not attempt a dialectical synthesis of the two philosophers; instead he directly raised Plotinian Neoplatonism to the highest level of Christian thought. Thus his notion of participation aims to provide a basis for the "catharsis" of the soul in its

---

[39] See K. Reinhardt, *Kosmos und Sympathie* (Munich, 1926), especially pp. 61ff.; M. Pohlenz, *Die Stoa* (Göttingen, 1948), I, especially pp. 208ff.

[40] Ph. Merlan, *From Platonism to Neoplatonism* (The Hague, 1953), p. 169. In his conclusion the author writes: "It is perfectly legitimate to speak of an *Aristoteles Neoplatonicus*." *Ibid.*, p. 195.

approach to the divine Ideas contained in the eternal Word "the vision of which [the ideas] makes the soul extremely happy." [41] The Neoplatonic derivations of the Greek Fathers (Basil, Gregory Nazianzen, Gregory of Nyssa) seem to be more profound and complex, but they had little influence on Western thought, which came into contact with Proclus' line of dialectical participation mainly through the complex of the *Areopagitica* and the opuscule *De Causis*. This latter, after St. Thomas' analysis of it in his commentary, was identified as an Arabic compilation of Proclus' *Elementatio Theologica* (Στοιχείωσις θεολογική). One can perhaps call the theory developed in the 12th century by the famous School of Chartres through the mediation of prominent men like Boethius and Chalcidius, a "closed notion of participation." It is a form of Platonism with a cosmic-mystic background that is based mainly on the *Timaeus*.

Medieval Augustinianism had established, under the authority of the great Doctor, a synthesis of elements derived from very different sources but tenaciously connected by a jealous tradition that showed its preference for Plato over Aristotle. In metaphysics it admitted, first, some sort of actuality in primary matter; later, it identified potency or receptivity with matter itself, so that matter became part of the essence of every creature: corporeal matter in corporeal creatures, spiritual matter in spiritual creatures (universal hylomorphism). Furthermore, since genus as an indeterminate logical element corresponds to matter which is the indeterminate ontological element, medieval Augustinianism admitted in every substance as many kinds of matter and form as there are logical genera and specific differences in its notion. For example, in man it admitted different kinds of matter and form for each of the following: substance, body, living, animal, rational, the individual Peter; all together, six kinds of matter and an equal number of forms (multiplicity of substantial forms). The methodological principle of this exaggerated realism consists in admitting a direct correspondence between logical and ontological order. Thus the genus is the matter while the difference is the form, and the parts of the definition are also the parts of the things themselves.

---

[41] St. Augustine, *83 Quaestiones*, q. 46, "De Ideis," PL 40, 29. See Hans Meyerhoff, "On the Platonism of St. Augustine's 'Quaestio de Ideis'," *The New Scholasticism*, XVI (1942), pp. 16–45.

St. Thomas, on the contrary, grasped from the very beginning the theoretical significance of the opposition between Plato and Aristotle and the absolute need to overcome it by bringing their fundamental principles and conclusions into agreement. This he did by elaborating his own notion of participation. This notion, in contrast with the Neoplatonic concordism, presents an entirely new concept and principle: it is the concept of *esse* as *actus essendi*, not to be confused with the *existentia* of Augustinianism and of rationalism. It is from the concept of *esse* as ground-laying first act that Thomas develops his own notion of participation and his entire metaphysics.

The first aspect of the Thomistic notion of participation is the Aristotelian concept of act as perfection *in se* and *per se*, and hence as ontological affirmation and positivity. Thus by its very nature act is prior to potency,[42] whether it is understood as operating activity,[43] or as form which is the first act (ἐντελέχεια) from which operation derives and to which it returns. In this view, although the corporeal essences are composed of two principles, matter and form, as "parts" of the essence,[44] yet in its metaphysical aspect essence gravitates toward the form which is the act.[45] Aquinas accepted this "primacy of act" without reservation and held the Judeo-Arab philosopher Ibn-Gebirol (Avicebron) directly responsible for the opposite view, namely, that the reality of beings is "resolved" into potency rather than into act against, therefore, both Plato and Aristotle. On the strength of his concept of act as prior to potency, Aquinas can demolish the fundamental principle of exaggerated realism. Genus and difference are concepts unified in the definition of the species and as such they cannot indicate distinct realities; rather, as expressive concepts, they both indicate simple formalities. Considered as parts of the definition that should itself be a unity of genus and difference, they indicate the same specific nature but in different ways. While the genus indicates the indeterminate element and the difference the determining element, the species indicates the determined whole of the synthesis.[46]

---

[42] *Metaph.* IX, 8, 1049 *b* 4ff.
[43] ἐνέργεια: *loc. cit.*, 1050 *a* 21–23.
[44] *Metaph.* VII, 10, 1034 *b* 2ff.
[45] *Metaph.* VII, 11, 1036 *b* 12.
[46] See *In VII Metaph.*, 1. 9, n. 1463.

Thus the logical composition of genus and difference does not by itself imply materiality; this latter must be shown in some other way, that is, from the data of experience. And even in material substances genus and difference as elements of the definition can be said to correspond to the matter and form of the concrete substance only indirectly or proportionally.[47] Genus, which is the indeterminate element of the definition, corresponds to matter, the purely potential principle; difference, which is the specifying element, corresponds to form, the actual principle. Moreover, *whenever it is known from some other source* that there are in existence substances which are absolutely spiritual, the genus and difference of their definition do not indicate any more two opposite ontological principles but rather the same formal reality considered first in its indeterminateness and then in the distinctive character of the individual spirits. Thus the angels or intelligences can also be said to be composed of genus and difference without thereby implying any composition of matter. For subsistent spiritual forms are endowed with the "power of understanding"—which can receive the universal intelligible forms without limits—and they receive act in exactly the opposite way (*per oppositam quamdam rationem*) as does matter, since this latter receives only individual forms.[48] The angels (and the human soul) are therefore simple in the essential order, inasmuch as they are subsistent forms *per se*, but they are composed in the entitative order, that is, of substance and accidents, of essence and the act of being. In this way Thomas has introduced *a new concept of both act and potency*. Whereas act is conceived simply as perfection or affirmation of *esse*, potency is conceived as capacity to receive perfection or as negation or privation. Two important consequences follow from this for Thomas: 1) Potency is not so named in one way alone, that is, simply as primary matter, but in just as many ways as there are of being subject to act. For potency is whatever takes on or conditions the act. Indeed potency, in addition to being primary matter, is also the human body in all its complexity: "Being a subject is not peculiar to the matter that is part of substance, but is a universal property of all potentiality."[49] 2) Primary matter, which is pure potentiality, is only

---

[47] *De ente et essentia*, c. 3; ed. Baur (Münster in W., 1926), p. 24, 6.
[48] *De spir. creat.*, a. I ad 2 and ad 24; ed. Keeler, pp. 13 and 19.
[49] *De subst. sep.*, c. 8; ed. Francis J. Lescoe (West Hartford, 1963), p. 76.

a subject and has no act whatsoever of its own. All actuality comes to it from form, so much so that not even God can produce matter without form.[50]

From this new concept of act and potency follows the second aspect of Thomistic metaphysics, which is also the thesis that drew the most severe attack during Thomas' lifetime, namely, the doctrine of the *unicity of substantial form* in all bodies, including living beings and man himself with his spiritual soul. This amounts to saying that the spiritual soul is *per se* and immediately the only substantial form of the human composite. A plurality of "forms," even if ordained or subordinate to the spiritual soul, even the admission of merely one intermediary form (*forma corporeitatis*), would destroy the essential unity of man. The spiritual soul, as ultimate form, would be no more than a perfective and therefore accidental form. As for the theological difficulty which was the main cause of the controversy, namely, that, if one followed the Aristotelian theory, Christ's dead body when separated from the soul could no more be called the body of Christ except equivocally (*aequivoce*),[51] Thomas accepts the conclusion but sees no objection in it to Christian theology. For the dead body of Christ, although indeed separate from the soul, remained at all times hypostatically united to the divinity of the Word.[52] Thomas also believes that the same rational soul, as the one and only substantial form, confers upon man not only the characteristic of spirituality but also man's lower ontological perfections: "Thus we say that in man there is no substantial form other than his rational soul, and that because of it he is not only man, but also animal, living, a body, a substance, and a being." [53] Hence the intellective soul is virtually the lower forms, inasmuch as it contains within itself the sensitive and vegetative powers that operate by means of the body,[54] just as in Aristotle's view a superior geometrical figure, e.g., the square, contains an inferior one, the triangle. Essences are in fact "like numbers," which differ according to the addition or subtraction of unities.[55]

---

[50] *Quodl.* III, q. I, a. 1.
[51] See *De An.* II, 412 *b* 21.
[52] See *Quodl.* II, q. I, a. 1 and ad 1.
[53] *De spir. creat.*, a. 3; ed. Keeler, p. 44, 1ff.
[54] See *STh.* I, q. 76, arts. 3–5.
[55] See *De An.* II, 3, 414 *b* 28; *Metaph.* VII, 6, 1043 *b* 34.

This is the third aspect of the Thomistic metaphysics of act, where the personal individuality of the spiritual principle is defended against the principal thesis of Averroism. The refutation of this thesis involves, as it were, two distinct moments: 1) The *phenomenological moment*, which is self-consciousness understood as the individual awareness that everyone has of being himself, the individual John Doe, the one who understands, wills, loves, etc.: *hic homo intelligit*. The intellect (and the will) is an act, individual perfection, on which depend all other values of the individual as a man who is also a person. This self-consciousness is at the foundation of all human life, of the rights and duties of each one as an individual man. Hence the need of "founding" this fact on a metaphysical basis. 2) The *metaphysical moment*, inasmuch as the consciousness of understanding (second act)—which is an absolute "first" in the spiritual life—can be explained only by admitting that every single man is endowed with an individual spiritual soul (first act), which at one and the same time is the substantial form of the body and outranks the body by its spiritual activities. The second act, the act of understanding, can be attributed to the individual man inasmuch as the first act also belongs to him.[56] Evidently the spiritual soul is the substantial form of the body because it is the principle of the vegetative and sensitive functions and not because it is the principle of the intellective functions by which it rises above the body and is a *per se* subsistent form.[57] The positive immateriality of the act of understanding affords us proof of the absolute spirituality of the human soul. Being endowed with an operation of its own that transcends the body, the human soul is a self-subsistent form to which *esse* belongs directly, and not in conjunction with the composite as in the case of material forms, and the soul communicates this *esse* to the body. Thus the immortality of the soul is proved on strictly metaphysical grounds. If the human soul as spiritual form is immediately and *per se* the subject of the act of being, this act or *esse* belongs to it in a definitive and inseparable way: "*Esse* properly belongs to a form, which is act. . . . But it is impossible that a form be separated from itself; therefore it is impossible that a subsistent form cease to be."[58]

---

[56] It is the substantial principle of intellection. See *De unit. intell. c. Averroistas*, c. 3, 80; ed. Keeler, pp. 50ff.
[57] See *ibid.*, c. 3, 60; *ed. cit.*, p. 38.
[58] *STh*. I, q. 75, a. 6. See also *ibid.*, q. 50, a. 5.

INTENSIVE HERMENEUTICS

The fourth aspect of Thomistic metaphysics is the affirmation of real distinction in all creatures between essence and the act of being (*esse*), which is the end result of the new concept of act. Today this is considered to be the key to the entire Thomistic system. While in Thomas' early works this teaching shows a direct dependence on Avicenna's "extrinsic" metaphysics, in his more mature works the distinction is the result of a better understanding of the primacy of act through the notion of participation. This involves two distinct points: 1) Pure perfection (*perfectio separata*) can only be one, and *esse* is the first perfection and the act of all acts;[59] hence subsisting *esse* is only one and this is God, whose essence is to be. 2) All creatures are beings by participation, inasmuch as their essence participates in the *esse* which is the ultimate act of all reality; hence the essence of creatures is related to *esse* as potency is to act.[60]

It is with this notion of participation that Thomas can overcome Augustinianism. He shows in fact that the soul and created intelligences (angels), although simple as far as their essence is concerned, are composed as creatures in the order of being. Likewise, by using this same notion of participation and the consequent metaphysical composition, he can vindicate against Averroism the absolute dependence of the intelligences on God through creation and conservation.[61]

---

[59] *De Ver.*, q. II, a. 3; *STh.* I, q. 4, a. 1 ad 3 and ad 2. The principle of *perfectio pura* is a dominant theme in *Liber de Causis* which was also known as *Liber Aristotelis de expositione bonitatis purae*. Cf. O. Bardenhewer, *Die pseudo-aristotelische Schrift 'Ueber das reine Gute' bekannt unter dem Namen 'Liber de Causis'* (Freiburg i. Br., 1882), p. 163. Thomas develops the dialectic of *perfectio separata* when dealing with the composition of *essentia* and *esse*. See for example *C. Gent.* II, 52; *De spir. creat.* 1, and especially his comment on proposition IV of *Liber de Causis* (Lectures 4 and 5).

[60] "It is indeed necessary that every simple, subsisting substance be either its own *esse* or participate in *esse*. But a simple substance that is subsisting *esse* itself cannot be but one, just as whiteness, if it were subsisting, could only be one. Therefore every substance that comes after the first simple substance participates in *esse*. But whatever participates [in another] is composed of that which participates and that which is participated in, and that which participates is in potency to that which is participated in. Hence every substance, no matter how simple it may be, if it comes after the first simple substance, is in potency to be (*est potentia essendi*)." *In VIII Physic.*, 1. 21; ed. Pirotta (Naples, 1953), No. 2491. This is a well-known text in the Averroistic school.

[61] *STh.* I, q. 44, a. 1 and ad 1; *ibid.*, qq. 104 and 105.

Finally, the notion of participation provides the formula for the analogy between creatures and their Creator: "Creatures are said to resemble God, not by sharing a form of the same specific or generic type, but only analogically, inasmuch as God is being by his very essence, and other things [are only beings] by participation." [62] In this latest conception of Aquinas *esse* is no longer the *accidens* of Avicenna, but rather the immanent act of the substance or *esse substantiale*, which is the proper effect of divine causality.[63]

A twofold consequence can be derived from the Thomistic metaphysics of act. First, the multiplication of the individuals within the same species (predicamental participation) can be explained through the principle of individuation, which is found in the potential part of the essence as determined by its "corporeity" (*materia signata quantitate*). Second, the doctrine of the principle of subsistence of beings by participation can be referred to *esse* as to the "actus substantiae": "Properly speaking, *esse* . . . is only attributed to the substance that subsists by itself."[64] The purely spiritual substances, i.e., the intelligences according to philosophers, the angels according to theology, are each one their entire species, for they lack the principle of individual multiplication which is matter.

In his later works Thomas fully adopted the thesis of Simplicius that there is a fundamental agreement between Plato and Aristotle.[65] This agreement is born out of the notion of participation, which provides the ultimate basis for the theory of act and potency and thus overcomes the obstacle of Greek dualism. The Thomistic synthesis is absolutely original: it accepts the metaphysical nucleus of Platonic transcendence (notion of creation, composition of *esse* and essence, doctrine of analogy) and welds it with the act of Aris-

---

[62] *STh*. I, q. 4, a. 3 ad 3. Fundamental analogy is therefore strictly metaphysical and expresses a constitutive and grounding relationship of *entia* to the *Ipsum esse*; it is the analogy of *plurium ad unum*. See Fabro, *Participation et causalité*, pp. 597ff; It. ed., pp. 496ff. For the foundation and division of analogy see pp. 607ff. In *La doctrine de l'analogie de l'être d'après Saint Thomas d'Aquin* (Louvain-Paris, 1963) B. Montagnes also emphasizes the theology of intrinsic attribution against the traditional interpretation of Cardinal Cajetan. See "Compte Rendu" by C. Fabro in *Bulletin Thomiste*, XI (1964), pp. 193–204.

[63] *Quodl*. XII, q. V, a. 5.

[64] *Quodl*. IX, q. II, a. 3 and ad 2: "*Esse* is that in which the unity of the supposit is grounded."

[65] See *De subt. sep.*, c. 3; ed. Lescoe, pp. 51ff. and 54ff.

totelian immanence (the unity of the substantial form, the intellective soul as substantial form of the body, the doctrine of abstraction).

The originality of the dialectic of *esse* was fully developed by Aquinas especially in his later works because of a more direct knowledge of some Neoplatonic writings, such as those of Proclus and Porphyry. In *De substantiis separatis* (1272–73) he solves the classic Neoplatonic problem of the accord between Plato and Aristotle through his "new" notion of participation. This he does in Chapter 3, *De convenientia positionum Aristotelis et Platonis*, where he shows the agreement of the two philosophers on such important doctrines as the real composition of essence and *esse* in creatures, the absolute immortality of spiritual substances, and the notion of divine Providence, while in Chapter 4, *De differentia dictarum positionum Aristotelis et Platonis*, he shows that their differences concern only very secondary points.

V

From the preceding discussion it becomes clear that in Thomistic speculation the notion of participation expresses the ultimate point of reference both from the static viewpoint of the creature's structure and from the dynamic viewpoint of its dependence on God. This notion takes from Platonism the idea of exemplar relationship and absolute distinction between participating being and *Esse subsistens*, and from Aristotelianism the principle of real composition and real causality at every level of participated, finite being. To assert, as has been done (Geiger), that Thomas holds as distinct participation by similitude (*secundum similitudinem*) and participation by composition (*secundum compositionem*), is to break the Thomistic synthesis at its center, which is the assimilation and mutual subordination of the couplets of act-potency and *participatum-participans* in the emergence of the new concept of *esse*. Such a view compromises at its root the meaning and function of radical *Diremtion* of the distinction between *essentia* and *esse*.[66] In the light of the new concept of *esse* the couplet *ens per participationem* and *ens per essentiam* takes on a more radical heuristic value with respect to the couplet of act and potency. But since the notion of *esse* as act of

---

[66] See Fabro, *La nozione metafisica di partecipazione*, pp. 12ff.; *Participation et causalité*, especially pp. 63ff.; It. ed., pp. 58ff.

all acts and of the forms themselves (*"actus" omnium actuum et ipsarum formarum*) was not known to Aristotle and is of Parmenidean-Platonic derivation, the Thomistic notion of participation constitutes in a most intensive sense (Hegelian) the *Aufhebung* of the opposition between Plato and Aristotle. Thus the authentic notion of Thomistic participation calls for distinguishing *esse* as act not only from essence which is its potency, but also from existence which is the *fact* of being and hence a "result" rather than a metaphysical principle. This will explain the confusion of those Thomists who, following Kant's apriori,[67] ground both experience and apprehension of being in the act of judgment (Maréchal, Lotz, Rahner, Metz), and of those, too, who speak in rather equivocal terms of a "Thomistic existentialism."[68] We must therefore stay with the notion of participation and avoid all equivocations and confusions: "I believe that the participationist motifs unquestionably present in the thought of

---

[67] Kant, *Kritik der reinen Vernunft*, A 599.

[68] See the substantial criticism of B. Lakebrink in his work, *Klassische Metaphysik: Eine Auseinandersetzung mit existentialen Anthropozentrik* (Freiburg i. Br., 1967), especially pp. 17, 28ff., 58ff., 127, 140, 262ff. The defenders of the transcendental interpretation of St. Thomas, who maintain that judgment is the first and only way of grasping the *esse*, rather than direct apprehension as grasping the fact itself of existing, can do so only by forcing the texts. "Saint Thomas never clearly says that the *esse* 'comprehended' by the act of judging is the unique, distinct, ontologically ultimate act of a thing rather than (or as well as) its facticity. He never affirms that our knowledge of the character of the metaphysical act of existing is grounded in our judgmental apprehension of it." Cf. G. Lindbeck, "Participation and Existence in Interpretation of St. Thomas," *Franciscan Studies*, XVII (1957), p. 22. In fact by deriving the *actus essendi*, understood in the nominalistic sense of existence, from the act of judgment, the neoscholasticism of Maréchal and his school (Lotz, Rahner, Marc Coreth, Brugger, Metz) has accepted the modern principle of immanence and attempts to introduce the "principle of the transcendental" into both dogmatic and moral theology. For Rahner's peculiar interpretation of Thomistic texts and contexts see C. Fabro, *Karl Rahner e l'ermeneutica tomistica* (Piacenza, 1972—a new edition is in preparation). While Rahner takes his inspiration directly from the thematic of Heidegger's existential Kantianism, B. Lonergan (cf. *Insight, Verbum*) accepts his "transcendental" directly from the *Kritik der reinen Vernunft*. See G. Sala, *Das Apriori in der menschlichen Erkenntnis*, "Studien über Kants Kritik der reinen Vernunft und Lonergan Insight" (Meisenheim am Glan, 1971), especially pp. 330ff. For the ambiguous nature of such a trend see James B. Reichmann, S. J., "The Transcendental Method and the Psychogenesis of Being," *The Thomist*, XXXII (1968), pp. 449–508. See also the review of this study in *Rassegna di letteratura tomistica* (Naples, 1971), pp. 145f.

Aquinas are a more likely source for the metaphysical theory of *actus essendi* than in the judgmental knowledge of existence emphasized by Gilson." [69]

The first and most fundamental division of participation is into *transcendental* and *predicamental*. The former is concerned with *esse*, with the pure perfections that are directly grounded in it; the latter is concerned with univocal formalities, such as genera with respect to species and species with respect to individuals. The former does not seem to present any particular difficulty, since it expresses the proper and principal meaning of participation. But the latter, too, is absolutely clear and indispensable for Thomistic philosophy, if we keep in mind the following five observations: 1) As far as their ontological content is concerned, genera and species are present in their respective subjects and must therefore be predicated essentially (*secundum* [*per*] *essentiam*) and not by participation (*per participationem*). This is the element of the Aristotelian doctrine of immanence. 2) With regard to the *mode of being* (and therefore the mode of being actualized in concrete reality), a genus is differently actualized in the various species according to different degrees of perfection, so that ". . . among the species of one genus one is naturally prior to and more perfect than the other." [70] 3) Thus participation involves no doubt a univocal nature or essence but

---

[69] See Lindbeck, *art. cit.*, p. 102. The progress of Neothomism from the controversial gnoseological questions—especially promoted by the Louvain school (Mercier and Noël in the Thomistic sense; Maréchal and his followers in the Kantian sense)—to the foundation of Thomistic metaphysics on the notion of participation, is the object of a study by Hellen J. John. See her article, "The Emergence of the Act of Existing in Recent Thomism," in *International Philosophical Quarterly*, II (1962), pp. 595–620. Referring to the thesis defended by Geiger (*La participation dans la philosophie de S. Thomas* [2d ed.; Paris, 1942]), the author observes that "his technical elaboration of the subject in terms of two distinct systems of participation, participation by composition and participation by similitude, gave less occasion for attention to the originality of St. Thomas' conception of *esse*" (*ibid.*, p. 611). The author's observation bears particularly on the real distinction between *essentia* and *esse*, which expresses the most profound and characteristic aspect of Thomistic metaphysics and to which Geiger has given little consideration.

[70] *Quodl.* II, q. III, a.6. This doctrine is of Aristotelian origin. See *Metaph.* X, 4, 1055 *b* 25; *Phys.* I, 4, 189 *a* 3. See its development in Fabro, *La nozione metafisica di partecipazione*, pp. 161ff.

only insofar as this is raised to a metaphysical level and considered as a "whole," that is, as a complex of virtual perfections that are being divided into or participated by the various species (for the genus) and the many individuals (for the species); otherwise the metaphysical foundation for real multiplication would be missing. 4) Thomas has always kept the two kinds of participation in close relationship: "To participate is like taking a part, and so when something receives a part of what belongs to another fully, it is said to participate in it. Thus man is said to participate in animality because he does not possess animality in its fullest sense. Similarly, Socrates participates in humanity, subject participates in accident and matter in form." [71] In this text the examples refer primarily to predicamental participation. 5) Thomas explicitly admits univocal predicamental participation in the proper sense of the term: "Whatever is predicated univocally of many things through participation, belongs to each of the things of which it is predicated, for the species is said to participate in the genus and the individual in the species. But nothing is said of God by participation, for whatever is participated is determined by the mode of the participant, and is thus possessed in a partial way and not according to every mode of perfection." [72] Without predicamental predication there would be no ultimate reason, properly speaking, for the multiplication of either the species within a genus or of the individuals within a species, inasmuch as multiplication in any order implies a difference and a differentiation, and hence a different kind of participation in a particular formality or act: "Just as this individual man participates in human nature, so every created being participates, if I may say so, in the nature of being (*naturam essendi*), for God alone is his own *esse*." [73] Just as static or structural transcendental participation is the composition of act and potency in terms of *esse*, that is, the real distinction between essence and *esse*, so static predicamental participation is the composition of act and potency within the sphere of essence, that is, the real distinction between matter and form in the

---

[71] *In Boeth. de Hebd.*, 1. 2, n. 24; ed. Taur., p. 396 *b*.
[72] *C. Gent.* I, 32 Amplius².
[73] *STh.* I, q. 45, a. 5 ad 1. It is worth noting that predicamental participation is placed in the prostasis, whereas transcendental predication is found in the apodosis.

material world and between substance and accidents in the order of finite being in general.[74]

Parallel to the division of static participation in its structural framework and dependent on it, is the division of dynamic participation as causality, inasmuch as being by participation stems from the being that exists by its very nature (*esse per essentiam*). There is then, in the first place, causal participation, which is the production of the common *esse* of all creatures by creation. Aquinas has thus reversed the principle of emanationism, *ab uno nisi unum*, which, as he himself observes,[75] amounts to the theoretical justification of polytheism. It is beyond doubt that the transcendental aspect of creation affects the whole finite being in its actual reality, its essence as well as its *esse*: "By bestowing *esse*, God produces also that which receives the *esse*, and thus there is no need for him to operate out of something that was already in existence."[76] However, from the viewpoint of its transcendental foundation, the position of essence is quite different from that of *esse*, even though both of them come from nothing through the same creative act of God: "From the very fact that *esse* is attributed to a quiddity, not only *esse*, but the quiddity itself is said to be created, for before having *esse*, [the quiddity] is nothing except perhaps in the mind of the creator, where it is not a creature but the creative essence."[77] The metaphysical

---

[74] For a complete exposition of the two kinds of participation in the sense we have explained cf. *Quodl.* II, q. II, a. 3. See also *De ente*, c. 6; ed. Baur. p. 47, 1: "From this it follows that he [God] is not in a genus, for whatever is in a genus must have a quiddity other than its act of being (*esse*), since the quiddity or nature of a genus or species does not differ by reason of its nature in the beings of which it is the genus or species, whereas the act of being (*esse*) is different in different things." For other texts cf. A. Pattin, *De Verhouding tussen Zijn en Wezenheid en de transcendentale Relatie in de 2ᵉ Helft der XIIIᵉ Eeuw* (Brussels, 1955), pp. 25ff, pp. 27ff.

[75] See *In l. De Causis*, 1. 3; ed. Saffrey 18, 14ff.

[76] *De pot.*, q. III, a. 1 ad 17. For an analogous situation of matter with respect to form see *ibid.*, a. 5 ad 3: "This argument [*Objection* 3: "Every action terminates in an act. . ."] proves that prime matter is not created *per se*; but from this it does not follow that it is not created under a form, for it is thus that it has actual being."

[77] *De pot.*, q. III, a. 5 ad 2. A parallel text, but in an inverted order, is to be found in *De Ver.*, q. XXI, a. 5 ad 5: "A creature is from God not only in its essence but also in its act of being (*esse*), which constitutes the chief characteristic of substantial goodness; and also in its additional perfections, which constitute its absolute goodness. These are not the essence of the thing."

status of essence is therefore subordinated to *esse*:

> An essence is called good in the same way as it is called a being. Hence, just as it has *esse* by participation, so it is good by participation. *Esse* and good taken in general are simpler than essence because they are more general, since they are said not only of essence but also of what subsists by reason of the essence and even of accidents themselves.[78]

That this is the relationship between essence and *esse* in Thomistic philosophy seems to be beyond question: created essences stem from the divine essence through divine Ideas, and this derivation is formally by way of exemplarity. Furthermore every essence, although an act in the formal order, is created as potency to be actualized by the participated *esse* which it receives, so that its actuality is "mediated" through the *esse*. *Esse* is the act that constitutes the proper terminus of transcendent causality (creation, conservation) and it is by virtue of this direct causality of *esse* that God operates immediately in every agent.[79] Hence the derivation of participated *esse* from the *esse per essentiam* is direct, and along strict metaphysical lines, as grounded act from grounding Act. In fact, the participated *actus essendi*, precisely as participated, is intrinsically dependent on God. But once it has been created, and as long as it is not being annihilated, it remains an actuated act to the full extent of its metaphysical import. It belongs therefore to God to be the cause of *esse* by virtue of his very nature.

It should be clear at this point that causality as transcendental participation extends to both creation and conservation,[80] to form as well as to matter.[81] While Platonism thought of matter as *non-ens*, Thomas teaches that matter participates in form and hence also in *ens*.

Causality as predicamental participation, on the other hand, is concerned with *fieri*, which is the becoming or development of created reality within the order of genera and species. Here is where obtains the principle of Thomistic metaphysics "form bestows *esse*" (*forma dat esse*), which seems to invert the causal relationship existing in the transcendental order. Form, as is well known, is the proper act of

---

[78] *De Ver.*, q. XXI, a. 5 ad 6.
[79] *STh.* I, q. 8, a. 1.
[80] *Ibid.*, q. 104, a. 1.
[81] *Ibid.*, q. 45, a. 4 ad 3.

the essence of finite things by which they acquire their degree of reality and perfection and are thereby disposed to receive the *actus essendi*. This principle, which is obviously of Aristotelian origin, has therefore a twofold meaning. In the first place, form bestows formal or specific *esse*, inasmuch as it is the constitutive act of every real essence either by itself, as in simple substances, or in conjunction with matter, as in material bodies. This is true whether the form is understood as form of the part, e.g., the soul, or form of the whole, e.g., humanity. In this sense the principle obtains that form bestows formal *esse* or, stated otherwise, it is the form that constitutes predicamental being in its ontological order, that is, in its real order, since it is essence that confers reality, while form is the principle *quo*, determining the essence: "The form is indeed compared to *esse* itself as light is to the act of illuminating or whiteness to the actuality of being white" (*comparatur enim forma ad ipsum esse sicut lux ad lucere, vel albedo ad album esse*).

Taken in its second meaning, which is more proper and is closely connected with the first one, form is said to bestow *esse*, inasmuch as it is only the real essence, determined by form as by its formal act, that is the true subject of the *actus essendi*:

> Further, *ipsum esse* is compared even to form itself as act. For in things composed of matter and form, form is said to be the principle of being in that it is the complement of the substance whose act is *ipsum esse*, just as transparency is in relation to air the principle of illumination in that it makes air the proper subject of light.[82]

Clearly, then, form is the true cause of *esse* but only within its order, inasmuch as it is the predicamental mediator between created finite being and the *esse per essentiam*, which is the First Cause.[83]

From this it is not difficult to make a step further and look into the crucial and so much debated problem of the "concurrence" of divine causality with human freedom. Created will, as secondary

---

[82] *C. Gent.* II, 54: *tertio*.

[83] See Fabro, *Participation et causalité*, pp. 344ff.; It. ed., pp. 330ff., 470ff. In a pertinent, although somewhat isolated and accidental, reference to St. Thomas, Heidegger reduces the "presence of God" in things to God's causality (*Nietzsche* [Pfulligen, 1961], II, p. 415). Causality is indeed the foundation for our knowledge and affirmation of the presence of God, but it is through creation that God is present precisely *per praesentian* "inasmuch as all things are bare and open to his eyes" (*STh.* I, q. 8, a. 3). Heidegger does not mention this kind of presence.

cause, is truly the total principal cause of choice, but this presupposes the influence of the First Cause which is even more so the principal and total cause in its own order. This is the transcendental order that embraces, from its very foundation, the being of a creature and the operation that is grounded in it. The following statement is enlightening: "Since the form of a thing is within the thing, and all the more so as it is prior and more universal, and God himself is properly the cause of universal being which is innermost in all things, it follows that God operates initimately in all things.[84] Properly speaking, the relation in question does not consist merely in the fact that God concurs with created freedom, but rather in the realization that just as he founds created freedom in its *esse*, so he founds it in its operations. He does so by embracing, as it were, man's free will and the act proceeding from it in its totality, without thereby interfering with the action of the secondary cause which remains the total cause [85] in its own order.

VI

To the extent that participation allows one to conceive the created universe in the complexity of its natures as a reflection of divine ideas or exemplars, one may speak of participation by similitude (*per similitudinem*) in the transcendental order according to a relation of dependence of the finite on the Infinite. This has been expressed by Boethius in the following terms: "From these forms, which are beyond matter, stem the other forms that are in matter and make up the body. It is only improperly that we call forms those which exist in the bodies, since they are merely images. They are similar indeed to the forms that are not joined to matter." [86] And Pseudo-Dyonisius writes in this connection: "We call exemplars the substantial reasons of all things preexisting individually in the mind of God." [87]

---

[84] *STh.* I, q. 105, a. 5.
[85] "It is also apparent that the same effect is not attributed to a natural cause and to divine power in such a way that it is partly done by God, and partly by the natural agent; rather, it is wholly done by both, according to a different mode, just as the same effect is wholly attributed to the instrument and also wholly to the principal agent." *C. Gent.* III, 70. See Fabro, *Participation et causalité*, pp. 488ff; It. ed., pp. 424ff.
[86] *De Trinitate*, c. 2; PL 64, 1250 D; ed. H. F. Stewart and E. K. Rand (Cambridge, Mass., 1943), pp. 12, 51–56.
[87] *De div. nom.*, c. 5, par. 8; PG 3, 824 C.

In the predicamental order of finite beings there is also participation by similitude in virtue of their relations of perfection and causality, so that all beings in the universe seem to have some sort of universal affinity. This manifests itself in the form of an attraction or universal sympathy that they have for each other, inasmuch as inferior beings tend to approach superior ones as though they would aim to participate in their perfections. This ontological affinity, which orders the entire world into a cosmos, can be expressed as the principle of metaphysical continuity of beings, which Thomas has taken directly from Pseuso-Dionysius in the following formula: "Divine wisdom joins the highest of the lower to the lowest of the higher." [88] In virtue of this principle, all created knowledge can be seen in terms of participation. Thus while the angel knows himself by his essence which becomes immediately intelligible to him, he knows the nature of other things only by means of infused species, i.e., by participation.[89] Likewise, while angelic knowledge, despite the limitations due to a creature, resembles to a certain degree the direct and intuitive knowledge of the divine intellect, man knows by participation to the second power, if we may so speak, both from the objective and subjective points of view. From the objective point of view, inasmuch as he knows through species abstracted from matter, and from the subjective point of view, because of his twofold intellect, i.e., active and passive or possible. Thus whereas the angel is essentially an intelligent nature, man properly speaking is a rational being. It is only imperfectly that man can be called intelligent, and that because of his participation in a superior form of knowing: "What is found most perfectly in superior substances, in man is found to exist only imperfectly and, as it were, by participation. Yet this little amount [or perfection] is greater than all other things that are in man." [90] Evidence of the human mind's participation in the angelic intellect is found in the understanding of "first principles," which lies at the foundation of all knowledge, both speculative and practical. In the latter case, the understanding of the first principles is called synderesis.[91]

---

[88] *Ibid.*, c. 7, par. 3; PG 3, 872 B. See Proclus, *Element. Theol.*, Prop. 147; ed. Dodds (Oxford, 1933), p. 128.
[89] *STh.* I, q. 65, a. 1; *De subst. sep.*, c. 13.
[90] *In X Ethic.*, 1. II, n. 2110.
[91] *De Ver.*, q. XIV, a. 2.

First principles are like "seminal reasons" of all knowledge and virtue and their understanding is likened to a "divine seal" and a "spark of the soul."

> This virtue is fittingly called "spark," for just as a spark is a small flying particle of fire, so this virtue is a small participation of intelligence with respect to the intelligence that exists in an angel. Hence also the superior part of reason is called "spark," because it is the highest thing in a rational nature.[92]

The principles of synderesis constitute the natural law as a reflection of the eternal law of God in the human mind.

> The light of natural reason by which we discern what is good and what is evil, which is the function of the natural law, is nothing else than an imprint on us of the divine light. It is therefore evident that the natural law is nothing else than the rational creature's participation in the eternal law.[93]

There is still another kind of participation that concerns man's tendencies and is closely connected with the predicamental order. It consists in this, that the will, or man's appetite in general, is morally justified in the attainment of its particular objects according to the degree of its participation in reason. "The appetite is naturally correct with respect to something, just as it is with respect to the ultimate end, inasmuch as everyone by nature wants to be happy; but with regard to other things, the correctness of the appetite is caused by reason, inasmuch as the appetite *somehow participates* in reason." [94] From this it follows that moral virtue derives its character as virtue from participation in the intellectual virtue of prudence,[95] whereas vice participates in the vice of imprudence.[96] The will, once it is directed by the intellect, can direct the sensitive appetite, which is thus ultimately subject to and controlled by the directives of reason. This may explain why sensitive appetite is said to be rational by participation, and occasionally, when it follows the

---

[92] *In II Sent.*, d. 39, q. III, a. 1. See J. Mundhenk, "Die Begriffe der 'Teilhabe' und des 'Lichts'," in *Psychologie und Erkenntnislehre des Thomas von Aquin* (Würzburg, 1935), pp. 8ff.; H. Wilms, "De scintilla animae," *Angelicum*, XIV (1937), pp. 194–211.
[93] *STh.* I–II, q. 91, a. 2. See M. Grabmann, *Der göttliche Grund menschlicher Wahrheitserkenntnis nach Augustinus und Thomas von Aquin* (Münster i. W., 1924), pp. 53ff.
[94] *In III Sent.*, d. 35, q. I, a. 1, sol. IV.
[95] *STh.* II–II, q. 47, a. 5 ad 1.
[96] *Ibid.*, q. 53, a. 2.

lead of reason, is even called "will by participation." [97] Even the senses, to the extent that they are rooted in the substance of the spiritual soul and contribute to the act of the intellect, manifest in their own way an imperfect participation of the intellect: "We see that sense is for the sake of the intellect, and not the other way around. Sense, moreover, is a certain imperfect participation of the intellect; hence, according to its natural origin, it proceeds from the intellect as the imperfect from the perfect." [98] The highest among sense faculties is the cogitative power, which manifests an even greater participation of the intellect "because of a certain affinity and proximity to universal reason." [99]

Thus from the principle of metaphysical continuity and affinity emerges a conception of the world as an orderly solidarity of all beings.

> Natures which are ordained to one another are related to each other as contiguous bodies, the upper limit of the lower body being in contact with the lower limit of the higher one.[100]

This structure of the world according to the various degrees of participation proceeds in an ascending order. "Whereas bodies participate only in being, souls participate according to their nature in being and life, and intellect participates in being, life and intelligence." [101] In this metaphysical extension of the notion of participation all the constitutive relations of being are actualized, both with regard to structure and causality, up to their highest degree. This consists in the attainment of their ultimate goal, which is imitation and similarity in being, and most of all in the joint action of an inferior substance or faculty and a superior principle. Here the Dionysian principle of metaphysical continuity is integrated with

---

[97] *Ibid.*, III, q. 18, arts. 3 and 4. See also *De Malo*, q. VII, a. 6 ad 1. The doctrine owes its inspiration to Aristotle: *Eth. Nic.* I, 13, 1102 b 13. See R. Eucken, *Ueber die Methode und die Grundlagen der aristotelischen Ethik* (Frankfurt a. M., 1870), p. 19.

[98] *STh.* I, q. 77, a. 7.

[99] *Ibid.*, q. 78, a. 4 ad 5. See also *In II De anima*, I, 13, n. 397: "The sensitive power, at its highest, has some share in the intellectual power of man, in whom sensitivity is joined to intelligence." For this doctrine, which has almost been forgotten by tradition, see C. Fabro, *Percezione e pensiero* (2d ed; Brescia, 1962), pp. 198ff., and especially pp. 222ff., 234ff., 238ff.

[100] *De Ver.*, q. XVI, a. 1.

[101] *In l. De Causis*, 1. 19; ed. Saffrey 106, 11–13.

the Aristotelian concept of participation in the object of thought (μετάληψις τοῦ νοητοῦ),[102] whose highest development is shown in the order of grace. Indeed, it is because of its participation in the dignity of the spirit that the intellectual nature "attains to unity, in which the species of its nature somehow consists."[103] Thus, on the one hand, a spiritual essence can extend itself to all things and, on the other hand, it can ascend and reach out to the possession of God. "Only a rational creature is capable of God (*est capax Dei*), for it alone can know and love him explicitly,"[104] and consequently "only a rational creature is directly ordered to God."[105] To be sure, the object of human happiness by virtue of the unlimited opening or potentiality of the spirit can be no other than goodness itself or the good by its very nature. "Man's perfect happiness consists not in that which perfects the intellect by some participation, but in that which does so by its essence."[106] Such is the ultimate goal that a created intellect attempts to achieve in a supreme effort.

The first step in man's elevation to supernatural life is faith, which is but an initial, imperfect participation because of the lack of vision (*carentia visionis*) of its object, the divine essence. In faith, Thomas says, "that light [of vision] is not perfectly participated."[107] Faith, informed by charity, is strengthened and elevated by the infused gifts of the Holy Spirit, and especially by the gift of understanding, "which helps in some way to see limpidly and clearly those things that belong to faith."[108] But the highest participation and supreme achievement of human life is beatific vision. "The last and most complete participation of his [divine] goodness consists in the vision of his essence, by which we live together socially like friends (*convivimus socialiter quasi amici*), for it is in the sweetness [of that vision] that beatitude consists."[109] The constitutive element of supernatural life is sanctifying grace (*lumen gratiae*), which, in St. Peter's words,[110] is "the expression or participation of divine

---

[102] *Metaph.* XII, 7, 1072 *b* 20.
[103] *In III Sent.*, d. 16, q. I, a. 1 ad 3.
[104] *De Ver.*, q. XXII, a. 2 ad 5.
[105] *STh.* II–II, q. 2, a. 3.
[106] *Ibid.*, I–II, q. 3, a. 7.
[107] *De Ver.*, q. XIV, a. 1 ad 5.
[108] *In Isaiam*, c. II; ed. Parm., Vol. XIV, p. 475 b.
[109] *In III Sent.*, d. 19, q. I, a. 5, sol. I.
[110] *I Petr.* 3, 5.

goodness," [111] "the participated similitude of divine nature." [112] The constitutive participation of grace is extended to the supernatural virtues, especially the theological, and to the gifts of the Holy Spirit on account of which the soul is no more directed by reason but by another rule "which is divinity itself participated by man in his own way, so that he does not act any more according to human nature but as if he had been made God by participation." [113] The highest form of participation created by God is the personal union of the Word with the human nature of Christ. This has become the primary source of all participation in grace by believers,[114] inasmuch as the human nature of Christ is the close instrument of the divinity.

VII

Whereas Platonic vertical participation is actualized merely as imitation of the Idea and hence as a fall, as it were, into non-being and the phenomenon, the Aristotelian horizontal causality is like an endless repetition of universal essence in the singulars. The result is that both theories tend to emphasize formal univocity. In contrast, the Thomistic notion of participation, founded in *esse* as supreme intensive act, makes it possible to pass from finite to Infinite Being through analogical discourse, which has in participation its beginning, middle, and conclusion.

There is in the first place the *analogy of proportionality*, which considers beings, such as creature and creator, accidents and substance, from a static viewpoint, that is, from the viewpoint of the contents of reality which they actually possess (*in facto esse*). This is a purely logico-formal way of considering beings. It comes at the end of the speculative synthesis and, in the Thomistic conception of reality, is undoubtedly valid. The proper structure of a creature is that of a being by participation, whereas it belongs to God to be *esse* by nature or subsistent *esse*. A creature is a being by participation on a twofold level: in the transcendental order, inasmuch as it is a composite of essence and *esse*, and of nature and subsistence; in the predicamental order, because of its composition of substance and

---
[111] *STh.* I–II, q. 110, a. 2 ad 2.
[112] *Ibid.*, III, q. 62, a. 1.
[113] *In III Sent.*, d. 34, q. I, a. 3.
[114] *STh.* III, q. 19, a. 1.

accidents, and, in the case of corporeal substances, of matter and form. To the extent that a created substance is composed of essence and *esse*, it is as far removed from God, the *esse subsistens*, as it can possibly be, and in this respect the terms "creature" and "God" admit of no measure or comparison. But since the essence of a creature has also its own participated act of being (*actus essendi*), its actualization is not merely a relation of extrinsic dependence; rather, it is based on the act of *esse* in which it participates and which it preserves within itself and is the proper terminus of divine causality. We have here two proportional similarities which correspond to one another. Just as to Being *per essentiam* corresponds, as act, *Esse per essentiam*, so to being by participation corresponds, as act, participated *esse*. Similarly, while the principal mode of being, i.e., subsistence, belongs to substance, so the secondary mode of being, i.e., inherence rather than a proper *esse* or *actus essendi*, belongs to accidents. It is this static analogy of proportionality that is expressed in the tension of similarity-dissimilarity according to the Platonic view of the vertical "fall" of being, which at the same time is a multiform extension of the inexhaustible fullness of God. Furthermore, it is precisely through this static analogy of proportionality that beings obtain, within their order, the proper consistency of *esse*, since each being has its own essence which is actuated by its proper act of participated *esse*. This view is in sharp contrast with the metaphysical theories of Dionysian-Avicennian inspiration according to which God himself is the *esse* of all existing beings. For Aquinas, not otherwise than for Heidegger, the difference between Being and essents is founded on the *esse*, as intensive emergent act, that is diversely participated in by each being.[115]

---

[115] I fail to understand how the *intensive esse* of Aquinas could have been identified with Heidegger's *Dasein*, which is essentially a temporal presence (*In-der-Welt-sein*). "The Heideggerian *Dasein* resembles the Thomistic *esse* which is the object of metaphysics, not the Hegelian conceptual *esse extensivum* but the *esse intensivum* of Cornelius Fabro, the notion of existence as the *Thesaurus of the Perfection of Being*. The *Thesauric Being* grasped by an intellectual intuition is not the most impoverished being like the Hegelian Pure Being, but precisely the richest of being." Yet the author of this passage admits immediately afterwards that Heidegger proceeds within the framework of the modern *cogito*, which cannot be affirmed of Aquinas. See William E. Carlo, "Metaphysics, Problematic or Science: Methodology vs. Science," in *The Nature of Philosophical Inquiry*, Proceedings of the American Catholic Philosophical Association, XLI

In contrast to static analogy of proportionality, there is the dynamic *analogy of intrinsic attribution*. While the former expresses in its own way a relation of similarity, the latter expresses mainly a relation of foundation and dependence of beings on *esse*. The analogy of proportionality emphasizes, as it were, the Aristotelian aspect of the immanence of *esse* in beings; the analogy of attribution, on the other hand, stresses the Platonic aspect of radical dependence of participant beings on the pure perfection that is separate from them. Obviously the analogy of attribution, in the sense that has just been explained, is ground-laying with respect to the analogy of proportionality, for it seizes and expresses the *esse* of being in its emergence as participated act with regard to the unparticipated Act. In this sense it can be said that analogy of proportionality presupposes, and is based upon, analogy of attribution. This amounts to saying: 1) that *esse* does not belong to creature (*ens per participationem*) otherwise than by participation in the Creator (*Esse per essentiam*); 2) that *esse* does not belong to accident (*ens secundum quid*) otherwise than by participation in substance (*ens simpliciter*).

The intrinsic relation of the two kinds of analogy can perhaps be expressed in the following terms: "Everything is therefore called good from the divine goodness, as from the first exemplary, efficient, and final principle of all goodness. Nevertheless, everything is called good by reason of the likeness of the divine goodness belonging to it, which is formally its own goodness, whereby it is denominated good." [116] Thus analogy of attribution accomplishes the ultimate

---

(Washington, D. C., 1967), p. 137. This same thesis, which in our opinion is absolutely groundless, is developed by the author in his volume, *The Ultimate Reducibility of Essence to Existence in Existential Metaphysics* (The Hague, 1966). We only wish to point out that for Thomas the object of metaphysics is not *esse* but *ens* (cf. *In IV Metaph.*, 1. 1, nri 529ff.) as "id quod habet esse." For the author's other arbitrary textual interpretations see the book's review by C. Vansteenkiste in *Rassegna di letteratura tomistica* (Naples, 1969), pp. 117ff. While disputing with the present writer, Carlo affirms: "We have an intuition of *esse* in the existential judgment grasping the object of metaphysics" (*op. cit.*, p. 96). This, as we have previously stated, is the thesis commonly accepted by the followers of Maréchal (Lotz, Coreth, Brugger, Marc, Rahner, Keller, Puntel), who ground *ens* on the *esse* of the copula as referred to the judgment of existence, more or less on common ground with Kant, Wolff, and nominalistic scholasticism (cf. Fabro, *Participation et causalité*, pp. 280ff.; It. ed., pp. 603ff.). Yet even Carlo admits that for Thomas "*verum sequitur* esse rerum" (*op. cit.*, pp. 99f., n. 1.). What, then, is the rationale of his reasoning?

[116] *STh.* I, q. 6, a. 4.

"resolution" of metaphysical discourse by relating the many to the One, the diverse to the Identical, and the composed to the Simple. It is at the same time the answer to the problem of the Parmenidean One within the creationist theory. Through his notion of intensive *esse* and the consequent distinction between *esse* and essence in creatures, Thomas not only duly emphasizes the difference between *esse* and being, but he also succeeds in making God's presence in creatures more active and meaningful than in the panentheistic theories of Dionysius, Avicenna, Eckhart, Cusanus, Spinoza, and Hegel. Whereas in these latter theories God as Being is the Act as *Essence of essences,* in Thomas' view God as *Esse per essentiam* is the principle and actuating cause of *esse per participationem*, which is the proper, actuating act of every real essence. Accordingly, the metaphysical background of this kind of analogy includes two distinct aspects. The first and more evident is the causal derivation of participated being from the *Esse per essentiam* in which it is grounded, or the relation of dependence of the former on the latter; the second and more profound aspect is the presence of the *Esse per essentiam* in participated being in virtue of his total causality and his "coming down to" and "getting into," as it were, created being itself. The Thomistic formula, *per essentiam, per potentiam, per praesentiam*,[117] expresses most effectively, along with the presence of the Absolute in created beings, the highest degree of dependence of the finite on the Infinite.

With regard to analogy, it is not possible to distinguish, as has been done for static and dynamic participation, between predicamental and transcendental analogy except in a limited sense. This consists in limiting, if one wishes to do so, predicamental analogy to the relation between substance and accidents and reserving transcendental analogy for the relation of creatures to their creator. This becomes even clearer if we note that the phrase "predicamental analogy" seems to involve contradiction, since it is precisely the predicamental order that seems to constitute the sphere of formal univocity. Predicamental participation is therefore, strictly speaking, confined to univocity. The genus is actualized in the species by

---

[117] *Ibid.*, q. 8, arts. 1–4. See Fabro, *Participation et causalité*, pp. 509ff.; It. ed., pp. 470ff.

means of the specific difference. The individuals of the same species possess the same specific constitutive characteristics, and what distinguishes them is their individual notes. Here is where the principle of predicamental participation obtains: "Whatever is predicated univocally of many things through participation, belongs to each of the things of which it is predicated, for the species is said to participate in the genus and the individual in the species." [118] And from the formal point of view it must be said that here the *ratio* applies in equal fashion, for here there is "division of a genus into its species which equally share in the generic nature." This is in contrast with the analogous *ratio*, which is predicated of one primarily and of others in a secondary way (*secundum prius et posterius*).[119]

Yet, regardless of what has been said, this predicamental participation would not be true participation, which it really is, if it were not somehow founded upon, or related to, analogy of being. This obtains when the species, and especially the individuals, are considered no longer on a purely logical level but rather as modes of being in concrete reality. Thus, when seen from this metaphysical aspect of reality, Peter and Paul participate unequally in human nature, that is, each one shares humanity in his own way, inasmuch as each one, as previously seen, has a different *esse*.

> The reason for this is that, since two things must be considered in a being, namely, its nature or quiddity and its *esse*, there must be in all univocal things a community of nature but not of *esse*, for any one *esse* is only in one thing. Hence human nature is not in two men according to the same *esse*. Hence also whenever a form signified by name is *esse* itself, there can be no question of univocity, for even being is not predicated univocally.[120]

For this reason we have defended the analogy of being with respect to individual singulars as within its normal scope since being as such cannot be but individual and singular. Thus predicamental partici-

---

[118] *C. Gent.* I, 32 Amplius.
[119] See *STh.* I–II, q. 88, a. 1 ad 1. The ultimate *ratio* of the analogy between beings and *esse* is therefore metaphysical participation which throws light on the doctrine of the transcendentals and distinguishes it radically from the so-called modern transcendental. Cf. C. Fabro, "The Transcendentality of "Ens-Esse' and the Ground of Metaphysics," *International Philosophical Quarterly*, VI, (1966), pp. 389–427.
[120] *In I Sent.*, d. 35, q. I, a. 4. See Fabro, *La nozione metafisica di partecipazione*, pp. 168ff.

pation functions as an intermediary and a notional bond between formal univocity and real analogy. However one point must be made clear. The transition from the predicamental to the transcendental order is made solely through the intensive emergent *esse*, which is the only transcendental medium.

There are therefore three moments in the grounding of the truth of being upon the notion of participation, and they are all linked together: composition, causality, predication. All three are founded upon and related to *esse*, which binds them together as universal act.

The metaphysical determination of *esse* as *actus essendi* in the sense of act of all acts, is proper to Aquinas and constitutes the transcendental foundation of the metaphysics of participation. This has been discovered by the strictly metaphysical method of resolution or reduction (*per resolutionem* or *per reductionem*),[121] as Aquinas often calls it, of accidental predicamental acts to substantial form and of both accidental and substantial acts to the more profound substantial act which is *esse*. It has also been discovered by the method of the absolute reduction of the act of being by participation to the *esse per essentiam*.[122] This is a kind of transition, as well as upward movement, from the given to act and from the finite to the Infinite. The latter is no longer considered here as a given or as contained but rather as a giver and container, the act present in every act, the perfection of all perfections, and consequently as an invitation for man to direct his thought and aspirations toward the Absolute.

The principle of act, as Heidegger has rightly observed, is crucial in the development of Western thought. The answer to Heidegger's problem is found in Thomas, provided this is sought within his view of the ultimate metaphysical determination. For it must be admitted that the Thomistic notion of act, which culminates in *esse* as the *actus essendi*, is the synthesis of the Platonic ἰδέα and of the Aristotelian ἐντελέχεια ἐνέργεια with which it shares the characteristic

---

[121] "All things must be traced to one first principle . . . by which they are coordinated." *De Pot.*, q. III, a. 6.

[122] This is in contrast with Heidegger's *Sein des Seienden* which is intrinsically finite. See C. Fabro, "The Problem of Being and the Destiny of Man," *International Philosophical Quarterly*, I (1961), pp. 407–36. Reprinted in *Tomismo e pensiero moderno*, pp. 135–64.

of immanence. A creature has therefore its own participated *actus essendi* which enters into real composition with essence as its transcendental potency. God, on the other hand, is the *esse per essentiam* or separated *esse* which is both transcendent and grounding. The Platonic principle of the ἰδέα or separated perfection [123] holds true only with regard to the *esse* as the act of all acts and of all forms, which was unknown to Plato as well as to Aristotle. This principle of separated perfection is eminently of Platonic origin and must be integrated with the Aristotelian principle of the emergence of act. Both principles are indeed founded on the synthetic Thomistic principle of participation. But despite his general acceptance of the Platonic principle of separated perfection, Thomas follows Aristotle in rejecting its application to the forms as such and, going beyond Aristotle who does not know *esse* as act, applies it exclusively to *esse*. Thus *Esse ipsum* or *Esse subsistens* is God himself who is the first, immovable, and separated Principle situated, as it were, at the summit of eternity (*in arce aeternitatis*). Hence God, as pure *esse*, is the grounding Act that is ever present in all acts, the present that actuates every presence. Likewise, God is the first and total cause, and he is at once both transcendent and immanent. He is transcendent but in a way quite different from a Platonic idea; similarly he is immanent but in a manner different from that of the Aristotelian act. He is immanent in the sense that he is the actuating, grounding principle of being, and not merely something accidentally contained in it. He is transcendent as the emerging incomparable act that is beyond all space, time, and measurement, for he is all in himself and

---

[123] In this sense the principle of *perfectio separata* has a structural meaning in Thomistic philosophy (cf. *De ente*, c. 5: "*calor separatus*," Baur 42, 18–43, 6; *C. Gent.* I, 43: "*albedo per se existens*," Leon. minor 41 *b*; *De div. nom.*, c. V, 1. 1, n. 629: "*si esset albedo separata*"; *De Causis*, Prop. III: "*albedo simplex, si esset separata*"; ibid., Prop. IX: "*si albedo esset separata*," ed. Pera, nn. 80, 235; ed. Saffrey 22, 11 and 66, 3; *Quodl.* II, II, 3: "*lux separata*"; ibid. III, VIII, 20: "*albedo subsistens*"; *De subst. sep.*, c. VI: "*forma, si separata consideretur*," Perrier 150, par 43). The typical Platonic element in these examples consists in the *separatio*, that is, the consideration of the form-quality as *subsistens* and hence as something distinct and unique. This *separatio*, which Platonism attributed to the simple forms as such and Aristotelianism ascribed only to the intellective forms (νοῦς and intelligences), is resolved by Aquinas into the act of all acts that is *esse*.

all things are in him, from him and for him. In other words, he is the being whom all men have called God.[124]

Very appropriately then Scheler, as he reflects on the metaphysical grandeur of man in the writings of his maturity, refers to Aquinas, along with some other great geniuses like Plato, Dante, Nicholas of Cusa, Giordano Bruno, and Goethe, as representative of the idea that man concentrates in himself the whole of creation. Quoting Thomas, he writes: "The essences of all things are interwoven in man, in whom they find their solidarity. *Homo est quodammodo omnia*, as we read in St. Thomas." This is the ideal Renaissance man as microcosm. "To aspire to culture," continues Scheler, "means therefore to search with loving concern for an *ontic participation* and a communion with all that is *essentially cosmic* in nature and in history, and not merely to search for an accidental existence and a particular kind of being (*Sosein*). It means, as for Goethe's Faust, to want to be a 'microcosm'." Explaining further the meaning of this kind of participation, he writes:

> We say: to know is to establish a relation of being and, more specifically, a relation of being that presupposes the forms of being of the whole and of its parts. It is a relation of "participation" of an essent in the reality of another essent. . . . The *mens* or spirit (*Geist*) means for us the *x* or the plexus of the acts of the *knowing* subject by means of which such participation is possible.

And he adds immediately:

> The root of this *x*, the determining moment of the movement toward the completion of the acts that lead to a certain form of this participation, can only be the parti-*cipation* that transcends itself and its own being. This is what we call "love" (*Liebe*) in its most formal sense. Without a tendency in the essent to proceed by itself and enter into participation in another essent, there can, generally speaking, be no knowledge whatever. And I do not find any other name to express this tendency except "love" or dedication—the breaking of the limitations of one's own being and essence through love.[125]

This is what in Scheler's phenomenological terminology is called

---

[124] "And this is what everybody understands by God" (*I via*); "to which [cause] everyone gives the name 'God' " (*II via*); "this [necessary being] all men speak of as God" (*III via*); "and this [supreme being] we call God" (*IV via*); "and this [first intelligent being directing all things to their end] we call God" (*V via*). *STh.* I, q. 2, a. 3.

[125] M. Scheler, *Bildung und Wissen*, 2d ed. by Maria Scheler (Frankfurt a. M., 1947), pp. 6 and 24ff.

"love for the essential" (*Liebe zum Wesenhaften*)[126]: hence vital experience, contemplative enjoyment of the object in the immanence of the subject of the forms. Hence also the identity of essence and existence and the attempt to draw St. Thomas into scholastic essentialism to which, under a subverted form, modern thought has been prisoner. Scheler is the essentialist of intuitionist vitalism.

On the opposite side, but on the same formal plane, is Heidegger, who is to be credited, however, for having made the fundamental resolution (*resolutio in fundamentum*) of the vacuum created by essentialist thought, including Scheler's notion of "love for the essential." Heidegger's being of the essent (*Sein des Seienden*) has no ground or foundation other than the giving of itself as "presence of the present" (*Anwesenheit des Anwesenden*). This is a new kind of identity of essence and existence but in an inverted form, as compared to that of Scheler, for it shows a tendency toward the outside of a world that is given as pure history (*Sein-Zeit* and *Zeit-Sein*).[127] Instead of the participation that is Scheler's Platonic mediation of love, we have here the resolution-dissolution of the essent into the nothingness that is Scheler's "veil" of being. Heidegger's being is therefore in a continuous process of falling, and by this falling it repeats and renews unceasingly its own presence, which is but the world's giving of itself to man in an unlimited and meaningless revelation. Briefly, it is a being whose only limitation and meaning consist in its relationship to the definitive nothingness that is death (*Sein zum Tode*). By taking away being as presence definitively and totally, death shows the actual invasion of nothingness and the nonsense or rather the impossibility of any kind of participation, and

---

[126] See M. Scheler, "Vom Wesen der Philosophie," in *Vom Ewigen im Menschen* (2d ed.; Leipzig, 1923), I, p. 68; 4th ed., I, pp. 67f.

[127] This identity is found in a recent writing by Heidegger, *Zur Sachs des Denkens* (Tübingen, 1969), pp. 16ff. The attitude of some contemporary scholastic philosophers who, following the lead of Johannes B. Lotz, insist on the resemblance between Heidegger's *Sein* and the Thomistic *esse*, can only be explained in terms of their adherence to the scholastic essentialism of the plexus *essentia-existentia* and a consequent neglect of the Thomistic *esse* and the Thomistic notion of participation. See, for example, H. Meyer, *Martin Heidegger und Thomas von Aquin* (Paderborn, 1964) and the more cautious, although somewhat equivocal, study of B. Rioux, *L'être et la vérité chez Heidegger et Saint Thomas d'Aquin* (Paris, 1963), especially pp. 126ff.

consequently, of any kind of transcendence of Act as metaphysical principle and τέλος.

In conclusion, if we compare Thomas' notion of being to that of Heidegger, we see that their positions show the greatest affinity and the greatest diversity. They both agree in denouncing the neglect of *esse* and in their radical insistence on grounding beings on *esse* as the act of the essent (*Sein des Seienden*). There is, however, a fundamental difference in their understanding of *Sein*. Heidegger conceives *Sein* as the actuality of an essence that is the inverted *cogito*, a consciousness, namely, that is entirely expressed in the world as *Dasein* and goes beyond Hegel's principle, *Das Aeussere ist das Innere* and vice versa. There is indeed, in Heidegger's conception, no more distinction between external and internal, but just one way of seeing *Sein*, which is immersed in time, as Kant intended it to be. In contrast, Thomas understands being as the metaphysical act that is participated in every act, while it itself does not participate in any act. As such, it can serve as a point of departure for the speculation of him who, while himself existing in time, is searching for the Absolute beyond time and is seriously concerned with the problem of his ultimate destiny.

University of Perugia.

BIBLIOGRAPHY

José M. Artola. *Creacion y Participation.* Madrid, 1963.
H. Beck. *Der Akt-Character des Seins.* München, 1965.
J. De Finance. *Être et agir dans la philosophie de S. Thomas.* Paris, 1945.
C. Fabro. *La nozione metafisica di partecipazione.* 1st ed. Milano, 1939. 3d ed. Turin, 1963.
———. *Participation et causalité.* Paris-Louvain, 1960. It. ed. Turin, 1961. (The two editions contain notable differences in the arrangement of material and in contents; the Italian edition has also indices of terms, authors, sources, Thomistic texts, both quoted and referred to, and an analytic index.)
———. *Esegesi tomistica.* Rome, 1969.
———. *Tomismo e pensiero moderno.* Rome, 1969.
L. B. Geiger. *La participation dans la philosophie de S. Thomas.* Paris, 1942.
A. Hayen. *L'intentionnel dans la philosophie de S. Thomas.* Bruxelles-Paris, 1942.
K. Kremer. *Die neuplatonische Seinsphilosophie und ihre Wirkung auf Thomas von Aquin.* Studien zur Problemgeschichte der antiken und mittelalterliche Philosophie, I. Leiden, 1966. (Cf. "Compte Rendu" by C. Fabro in *The New Scholasticism*, XLIV (1970), pp. 69–100.)
K. Krenn. *Vermittlung und Differenz.* Vom Sinn des seins in der Befindlichkeit der Partizipation beim hl. Thomas von Aquin. Rome, 1962.

G. Lindbeck. "Participation and Existence in the Interpretation of St. Thomas Aquinas." *Franciscan Studies*, XVII (1957), pp. 1–22, 107–125.
B. Montagnes. *La doctrine de l'analogie de l'être d'après Saint Thomas d'Aquin.* Louvain-Paris, 1963. (Cf. "Compte Rendu" by C. Fabro in *Bulletin Thomiste*, XI (1964), pp. 193–204.
Io. Mundhenk. *Die Begriffe der "Teilhabe" und des "Lichts" in der Psychologie und Erkenntnislehre des Thomas von Aquin.* Würzburg, 1935.
F. Ocáriz. *Hijos de Dios en Cristo: Introducción a una teologia de la participación sobrenatural.* Pamplona, 1972.
L. Oeing-Hanhoff. *Ens et bonum convertuntur.* Münster i. W., 1953.
Carl J. Peter. *Participated Eternity in the Vision of God: A Study of the Opinion of Thomas Aquinas and His Commentators on the Duration of the Acts of Glory.* Rome, 1964.
E. Scheller. *Das Priestertum Christi.* Paderborn, 1934. Especially pp. 59ff., "Quellen der Anteilnahme bei Thomas."
G. Siewert. *Der Thomismus als Identitatssystem.* Frankfurt a. M., 1939.
G. Söhngen. *Thomas von Aquin über Teilhabe dürch Berührung,* in "Die Einheit der Theologie." München, 1952.
F. Ulrich. *Homo Abyssus: Das Wagnis der Seinsfrage.* Einsiedeln, 1961.
Various Authors. *De Thomistische Participatieleer.* "Studa Catholica." Nijmegen, 1944.

ENRICHING THE COPULA
KENNETH L. SCHMITZ

It is a commonplace among students of St. Thomas Aquinas that in his view a judgment does not come to rest in its truth until it reaches the thing being judged about. For him the judgment in its fundamental nature is not simply a union of subject and predicate, but is rather the surge of the mind itself towards rest in the being of things (*esse rerum*). The judgment terminates not in a mental construction but in the thing itself (*ad rem*). This has led many Thomists to salute the judgment as the cash-value of knowledge. Usually our attention is directed emphatically to what is said in a judgment, and therefore to the terms which express its first-blush meaning. Ordinary attention takes the copula of the judgment more or less for granted. Like a familiar shy companion it is seldom noticed and barely audible. The historical study of St. Thomas' writings in this century, and the Thomism which has grown up with that study, has shifted attention to the copula and urged its importance. Under the influence of an "existential" interpretation of St. Thomas the central disclosure of his thought is held to be the radical value of *esse* (to be, be-ing). It is not surprising, then, that some Thomists have placed the key to the truth of the judgment in the copula rather than in a correspondence between the terms and the reality signified by them. Accordingly, *adequatio* of mind with reality is brought about in and through the copula.[1]

I

The history of philosophy provides examples of attempts to vindicate the adequation of thought with being. Thought has

---

[1] I do not intend in this commemorative essay to contribute to the recovery of the sense of St. Thomas Aquinas' writings, although that continues to be an exacting and significant task. My purpose is neither to restate nor to continue his doctrine, but to find in it suggestions for another line of thought, a line which St. Thomas may never have intended even remotely, and which in the end may even be incompatible with his intention. To a student of St. Thomas the differences will be obvious. Nevertheless, I hope that my indebtedness will not remain too obscure.

sometimes armed itself in the Cartesian manner with criteria for measuring its own conformity with being. But such an immediate and direct appeal to "pure" thought rests inescapably upon a tacit appeal to a human experience which includes sensible factors; and so it begs the question. Moreover, it seems to me that all attempts fail which try to join the knower and the known by putting the idea or judgment between them. For whether the intervention is by criterion, instrument, or method, it merely complicates the problem of their union by adding a useless intermediary. Indeed, in the face of the idealism of Descartes and Locke, and within the presuppositions of a theory of noetic species, modern scholastics were right to insist upon the "non-entity" of ideas and judgments. They insisted that a noetic principle, such as an idea or judgment, is nothing in itself but a pure sign (*medium quo*), and that it is not an object which, while signifying something else, also exists in its own right as a mental entity (*medium quod*). In trying to account for the concrete unity of language and cognition—which I may be forgiven for calling "linguo-knowledge"—there is much to recommend a theory of signs which differentiates knowledge-signs from all other kinds. It is an attempt to do justice to the quasi-instrumental features of linguo-knowledge, while at the same time keeping before us its radical ontological openness. Nothing *intervenes* between knowing and being, least of all the "equipment" of knowing, its ideas and judgments. The relationship between language and knowledge, however, would have to be determined more precisely since linguistic signs lend themselves to being considered as odd sorts of instruments, and by some linguists even as entities. Nevertheless, although linguistic signs are inseparable *in concreto* from noetic signs, it remains true that noetic signs are non-entitative, for there is a fundamental sense in which knowing is not an affair of entities or instruments at all. The essential possibility of knowledge lies in what has traditionally been called its immateriality and even its spirituality.

In much of current philosophical speculation, however, the duplicity inherent in linguo-knowledge seems to have split up into two theories which are for the most part indifferent or even opposed to one another. The instrumental theory, instanced by its cybernetic versions, treats knowing as an affair of signals, codes and receptors. The revelatory theory of ontologists such as

Heidegger, on the other hand, considers knowing to be an opening or clearing. Now it seems to me that in its roots knowing is not instrumental and that scholastics and ontologists in their different ways are right on this score. Of course, the factors that constitute knowledge can be disengaged and looked at in themselves. Indeed, it is important to remark that such a disengagement is not practiced only by theorists, for it is done by knowing itself. Thus, a natural reflection upon the act of remembering what I had previously seen or said innocently posits "a something which remains for recall," the remembered image. This is almost inevitably taken to stand in place of the original perception, by a kind of substitution. Some theorists, coming upon this rudimentary separation of image from object, sharpen the self-subsistence of the various factors in order to analyze them abstractly in and for themselves. When, at the end of their analysis, they integrate them once again in their unity, they may easily re-shape them instead within a synthetic context brought about by their own analysis. The original unity of experience may easily be lost. The factors, a term or a proposition, may then become a sort of thing which operates in itself, and the whole account of knowing may be surrendered to a set of instrumentalities and entities. The task of an ontology of knowledge, on the contrary, is to recover the original integrity of knowledge and to reinstate the recognition of the non-instrumental openness of language and its thought to the world in which it is situated.

It is not enough, however, to vindicate the truth of the judgment by simply pointing to the coincidence of the copula with the thing affirmed or denied by it. This would replace vindication by mere indication. Hegel's objections to Kant may not reflect nuances important to the older philosopher, but they are an instructive objection in principle regarding this very point. Hegel considered the Kantian account of the synthetic a priori judgment to be deficient because it posited no more than a "happy coincidence" between subject and predicate, and therefore failed to establish subjective thought in objective being. Indeed, the synthetic a priori judgment eventually fell apart, according to Hegel, into empty analytic forms of thought (categories) and unjustified syntheses of sense (intuitions). The Kantian theory of judgment complacently let the copula merely *assert* the union of subject and predicate under the indeterminate concept of an object. The result was a bare mindless assertion without any title authenticated in terms of thought, and based merely upon inspection of the sensible manifold and the "lucky find" of a given content submissive to the forms of thought. Kant's radical dualism forbade a transformative principle which could authenticate the claims of experience in terms of thought. To be sure, he subordinated the sensible content to the a priori forms of thought, but he did it as a master imposes his own conditions upon a slave. And he got what a master gets from a slave, not his genuine being but something less, —an appearance.[2]

A Thomist presumably is not plagued by the radical dualism of Kant. Nevertheless, although it must avoid a Cartesian criteriology, a philosophy must provide more than a Kantian critique. For any philosophical account of judgment must recover the validity of the connective in terms of thought. And so, while a philosophy may recognize that real being is reached in and through the sensible, still it cannot rest content with an *immediate* concession to it, but must appeal to the sensible on grounds that also lie within thought itself. For thought, after all, must institute the sensible in its own thoughtful mode of being. St. Thomas provides an example of an attempt to meet this need with respect to the simple apprehension of universals. For he taught that, in abstracting the species in the phantasm, the agent intellect actualizes it in the possible intellect, and thereby proceeds to recover in thought that very meaning towards which the mind directs itself. Moreover, his theory of participative causality provides a principle of transformation common to both being and thought.

The transcendental turn and the empiricist demand have more recently brought new pressures to bear upon the efficacy of the judgment. The union of thought and being in the judgment, therefore, must be thought through in terms of these pressures. Failure to secure the grounds for the truth of the judgment in the face of these pressures threatens to impede the drive of the mind towards that objectivity in which thought seeks to become more itself

---

[2] G. W. F. Hegel, *Glauben und Wissen* (hrsg. v. G. Lasson, 1928), F. Meiner, Hamburg, 1962, pp. 1–40.

by letting the thing be more and more in its own being. There is here a striking similitude between cognition and the love of friendship, for both must let the other be in its appropriate being—the known in its, the friend in his—while sustaining that being in their own. But in going out to and reaching the thing, knowing in that very activity of assertion returns also into itself and becomes more itself. So that, in returning to itself, it seals its discovery with the transcendental completeness of truth.

It may seem that the judgment promises more than it can deliver. For the asserted statement is but one wedge in a larger intelligible pie, the act of asserting is but one attestation in a life of thought, and what is asserted to is but one item in an ontological network. Hegel's strictures upon the judgment, therefore, run deeper than his dissatisfaction with any previous theory of judgment. To him the weak point of the judgment is the undeveloped meaning of the copula, for it never attains to a fully articulated meaning. Under the traditional logical forms of judgment, in which a predicate is attributed to a subject, there lies according to Hegel an ontological process. It begins with immediate empirical judgments in which a predicate is simply posited as inhering in a subject. The copula is unenriched, an abstract "is," a bare assertion. More developed forms of judgment achieve greater universality and necessity until the most perfect form is reached in the apodeictic judgment. Thus, a movement of mutual interpenetration binds the terms through the integrative power of the copula. For the copula is the energy of the mediation. It is also, for Hegel, the reconciling presence of speculative reason and the promise of absolute spirit. Indeed, the whole story of the judgment for Hegel lies in the gradual "coming of age" of the copula. Still, it fails to reach full maturity in the judgment, for in its very fulfillment it passes over into the middle term of the syllogism in search of a more adequate and expanded vehicle of expression.[3] Undoubtedly the Kantian deficiency, as well as his own conception of the ultimate aims of philosophy, lies behind Hegel's depreciation of the copula. The judgment is, for him, a partial attempt

---

[3] G. W. F. Hegel, *Wissenschaft der Logik* (hrsg. v. G. Lasson, 1934²), F. Meiner, Hamburg, 1967, 2 Bde., II, 264–308. (English: *Hegel's Science of Logic* (tr. A. V. Miller), Humanities Press, New York, 1969, pp. 623–63.)

to reconcile the dualism of ought and is, of ideal and real, of fact and value. Without accepting his dual problematic, however, or his dissatisfaction over the allegedly abstract and indeterminate form of the copula, it is still possible to take seriously his suggestion that the copula must be understood through a process of concretion and fulfillment. Lacking such a process, a philosophical theory, whether criteriology or criticism, whether empiricism or realism, must leave the copula as an immediate and merely asserted conjunction without a ground in an intelligible structure, a mere complacency of the mind in the opportune "fit" of the judgment with actual states of being. Of course, Hegel is right that the judgment must pass over into and become part of a larger discourse, with other kinds of connectives. Before it does, however, more can be said on behalf of the copula than Hegel thought. For the copula shows itself to be the *actual* unity of thought and being. In the present essay its features are condensed into a theory, but they are the very power which moves speech from one statement to another.

II

Any attempt to "make metaphysics" out of the copula as the actual unity of thought and being must face charges of cultural and linguistic provincialism. It will be said that the copula is merely a logical-linguistic device, that many languages do nicely without one, and that the copula distributes the understanding of reality into subjects and attributes. This is undoubtedly true. It is also true that the full meaning of a philosophy is affected by the language within which it is conceived and expressed. Otherwise translations could be substitutions. No doubt, too, philosophies written in non-copulative or non-attributive languages would disclose different aspects and emphases about reality. To admit this, however, is not to concede to what is suggested, viz., that distinctions properly drawn in one language must be essentially incommunicable in another and even false in reality. There is nothing in the generalized attacks upon the particularity of a linguistic structure which shows why the distinctions drawn within it must be null and void. There is nothing in the attempts to discredit Greek and Indo-European languages generally that aborts distinctions carefully drawn in and through them. Thus, for example, the most sweeping attacks upon "Hellenistic" modes of thought seem to be promoted by an assumption that is as uncritical in the twentieth century as was a similar shock at the discovery of non-European cultures in the seventeenth. Both share the assumption that radical difference means error somewhere, and that differing conceptions render each other relative, subjective, and imaginary. Linguistic self-criticism is valuable when it is directed against one or another false extrapolation of a linguistic feature into a metaphysical structure. These failures, of course, are not strictly failures of language as such, but are rather the user's failure to understand the scope and import of the locution he uses. In extending the attack to a whole cast of mind and to an entire language or family of languages, however, the critique comes into danger of a self-referential inconsistency. For the more extreme "anti-Hellenists" argue in an attributive language to undermine the possibility of attributive modes of expression. The possibility, not only of philosophy, but of discourse itself comes into question, and with it thought itself. For philosophy requires some particular language, just because thought itself requires such a particularity. If we are forbidden to philosophize in some particular language—in our own, in German, Latin, Greek, or some other—then philosophy becomes impossible for us. And if this stricture were extended without discrimination to all languages in principle, then not only philosophy but thought itself would perish, including that of the "anti-Hellenists."

I shall make two suppositions about language, therefore, which at least some critics may not share with me. 1) Language as such, and therefore any language—even the Greek, it may be necessary to add—is an aid to thought, however uncertain. Language is not the enemy of thought. It is not its trap. 2) On the contrary, thought is the energy of language and outstrips the forms of language. Such transcendence is a quite general characteristic of other human activities, such as technology, but in language it proceeds in accordance with the heritage of a particular speech community. The transcendence of thought, however, does not mean that language is simply its external shell. It means rather that, while language is the embodiment of thought, the latter is not so identified with language that it simply *is* linguistic behavior. A thought entrapped by language would deserve its fate, for it would be dead thought, unable to enliven its own body. The expression of meaning demands an embodiment of thought in language in such a way that neither language nor thought remains indifferent to the other. Nor can the two together, linguo-thought, remain indifferent to the demands and possibilities of life. There arise in thought and speech, therefore, non-cognitive as well as cognitive modes of expression, and no theory of judgment can be adequate which disregards them. Nevertheless, the focus of judgment is cognition. The cognitive structure of language manifests itself in the sentences of language, the propositions of logic and the judgmental activities of thought. Of these three, the sentence is the most conditioned by the particularity of its language. The proposition is a formalization which comes under more universal conditions of thought. The judgment is the cognitive act by which thought, seeking its ends within human life, seeks them precisely in and through the sentence and the proposition.

In its cognitive aspirations, thought seeks to know things *as they are,* and so it is a transcendental necessity of thought that every language be *assertive.* The assertion that such and such is so and so may be either a descriptive or, more generally, a declarative utterance. The linguistic means available for such utterances will vary from one linguistic family to another. But the need to assert, to describe or to declare, and to judge is written into the nature of human thought, action, and life. The concrete demands of human life require all men to recognize and express the difference between saying and doing and between thinking and being. These distinctions are not exclusions, of course, for saying is a manner of doing, and thinking is a way of being. Still, there is an important difference between speech or thought informed by a presence other than itself and speech or thought that is empty of it,

[a mere] speech of the self that must sustain itself on speech.[4]

Speech is untruthful when it is empty, and false when it is misguided. These basic distinctions between saying and doing and between thinking and being are the fabric out of which the truth

---

[4] From "The Well Dressed Man With a Beard," by Wallace Stevens in *Poems* [ed. S. F. Morse], Vintage, New York, 1959, p. 104. I have taken the phrase, but not the full sense it has in the poem.

or falsity of assertions is cut, and each language must preserve them in its own way. Assertions may be brought about in some languages without a copula, but not without some intended copulation. The copulative function may be discharged without using an onto-existential family of verbs. In our own and cognate languages, however, the linguistic incarnation of the assertive force takes the form of a copula, and precisely some form of the verb "to be." Moreover, the reality which calls forth "is" and "is not" to service in the judgment also calls forth philosophy to an onto-existential reflection.

III

The first step in reducing the copula from its apparent brute immediacy is to distinguish the *esse* of the copula proper from the *esse* of the being in which the judgment terminates (*ad rem*). The former may be called a mental construct or being of reason (*ens rationis*) to distinguish it from a real being (*ens naturae*). The aim of the reduction, however, is not simply the disengagement of mental from real being and the separation of two modes of *esse*. Quite the contrary, it is a liberation which allows reflection to study the being of the copula and to unfold its capacity for bringing meaning to truth. The self-realization of thought is a very peculiar opening out on to the thing (*processus ad rem*). For it is both an advance of thought into being, and a trial to determine the objectivity of thought. Thus, the copula discloses itself as a linguistic function which is sponsored by thought in the name of the truth of what is being judged. It takes its sense from the being which, in being brought to judgment by thought, provides the norm for that judgment. This interchange constitutes the paradox of objectivity: that the judger judging is brought to judgment by the necessity of submitting his judgment to the terms of the one being judged. And so, the thing which utters no word determines the deciding word (*krisis*), or rather settles the appropriate range of words that can be spoken within the terms of the language. The primary distribution of the thing in terms of thought (*Ur-teil*) is a process which arises from the speaker who pronounces judgment, but it joins thought to being under the law of being (*jus, judicium*). The dictionary etymologies of both *res* (cf. reality) and *thing* are tantalizing, even if doubtful. *Res* may have been derived from *reor,* which means "to reckon, think, judge," making *res* that which is thought about, as in the term "public affairs" (*res publica*). So, too, *thing* may be the object of *think* (as *Ding* of *Denken*), making *thing* that which thought is about. In any event, both terms can indicate the correlative of thought, which is then a *processus ad rem.*

Nevertheless, it is easy to misunderstand the law of the thing (*jus rerum*), in accordance with which the decisive word is taken from the thing. We must, therefore, redress a possible imbalance by restoring the rights of thought (*jus rationis*), its freedom and integrity. For the decisive word, though spoken according to the thing, is uttered by the speaker in his judgment. To avoid misunderstanding the meaning intended by the term *thing* here, we must broaden and lighten the too restrictive and solid sense often given to the term today. A similar emendation must be made for other terms used to designate the focus of judgment. Such terms as "entity" and "being" need not exclusively designate a subsistent being, but may mean only that ultimate and decisive situation which is resolved by the judgment into terms of being and non-being. Thus a judgment may terminate in *anything:* in Hamlet as *dramatis persona,* just as well as in a natural thing; or it may terminate in a well-formed idea as a "good thing," or in our present purpose as a "thing of some difficulty;" or again, a judgment may refer to a lost childhood or a lost leg. Language and examples usually favor affirmative judgments which terminate in real things, the familiar trees and tables, but it must be remembered that negative as well as affirmative judgments can terminate in situations of existence, and that affirmative as well as negative judgments can terminate truthfully in situations of non-existence. Thus, the attributive form of a judgment which discloses a privation may be similar in its linguistic form to other judgments, but its full onto-existential intention is privative, for it does not simply deny a real childhood or a real leg, it marks the lost thing precisely *as lost.* The judgment, then, is the surge of the mind as it resolves something into its ontological character and mode, but the thing may be in an order of real beings, of fictions, projects or privations, or in some other manner or condition of being and non-being. Indeed, it is in a variety of ontological modes. When a guide refers to the solid thing before him as "Churchill's writing desk," he discloses an object constituted of past, present and future modes of being, including ones of fact and value.

It should be clear, then, that the thing in which the judgment terminates need not be static or even completed. Sometimes the thing being judged comes about because of the judgment, or even in and through the very judgment being made about it. Such a judgment does not terminate in an already finished thing, for it discloses not only the emergence of the thing but sometimes also the dependence of the thing upon the judgment itself. This is especially true of the practical judgments which swarm into being around our purposes, for performative utterances are the very stuff out of which personal lives and social institutions make further meaning. Such judgments are true if they disclose just what and how the thing is in its appropriate way of being. So great is the freedom of thought that it may even utter its protest in the face of the very being which it discloses in its judgment; but then it passes over into a judgment of value, for "it is" becomes "it ought." This continually happens with situations in social time and space which are still underway and which are partially constituted by the very judgments rendered upon the developing situation from within it. So too, the present essay addresses itself to a component of judgment, the copula, which itself comes into being only in and through the judgment. For the copula is a thing only in its coming alive in and through the judgment which employs it. Our present reflection, then, is about just such a strangely broad and unsolid *thing*. When we insist upon the law of the thing, we merely insist that human thought cannot confer truth simply and solely upon itself, but that it can come into its truth only by resolving the thing into its ontological situation by judging it in terms of the value of being and non-being. Thought submits itself to the full range of being and non-being in a way that preserves its own integrity by disengaging itself from being submerged by real subsistent being. It identifies the thing judged about within the context of being and non-being, but with all the shades and modes which various ontological situations embody, whether subsistent or insubstantial, finished or in process, real or ideal.

The distinction of the *esse* of the copula from the *esse* of the thing, then, is not meant to throw the judgment back upon purely mental structures (*entia rationis*). Indeed, the copula is just that which, in uniting the subject and predicate, intends the *esse* of the thing, and has no other justification for that conjunction than the *esse* of the thing. The distinction is made, rather, in order to break the unqualified identity of copulative and real being, and to replace it with an intended identification. Unless the distinction is made, the *esse* of the copula and the *esse* of the thing will remain immediate values whose coincidence may be accepted as a happy accident which cannot, however, be grounded intelligibly in the necessity of thought. This would be the defect of any philosophy which remained satisfied with merely asserting the confrontation of the mind with sensible being under the law of the thing and in an intellectual intuition. For such satisfaction confuses an assertion with an account. It recognizes the role of perception in knowledge and the law of the thing, but it provides no explanation of how the conjunction is possible. In order to ground the judgment intelligibly, that is, in the necessities of thought, we must give an account of it not only in terms of the law of the thing (*jus rerum*), but also in terms of the grounds for the possibility of truth (*jus rationis*). And this carries us further into the significance of the copula.

IV

In bringing about the union of thought and being the copula functions as a limit. Our initial sense of a limit is of that which stops and excludes, but a limit also joins. Moreover, it belongs to the beings limited by it. Does the copula belong to the knower? Yes, for it is a judgmental device, although we do not fall back through it upon an empty subjective thought. On the contrary, it thrusts us towards the objective of thought under the law of the thing. Does it belong, then, to the thing? Yes, for it discloses the thing in its being, although it remains an expression of linguo-thought.

Before we plunge into the ambiguity of the copula taken as a connecting limit, we ought to clarify three senses of the term *being* as it is used here. 1) The expression, "the thing in its being," means the thing resolved into its ontological situation. It designates the thing understood in terms of its mode or modes of being and its ontological relations with other beings, including knowers. 2) Since thought *is,* it too is a version of being, for it is the activity or process of opening out onto the thing and of returning to itself. Thought as active process *is* in some sense, even when it is mistaken; but 3) we may speak of thought coming into its truth as a coming into a fullness of being appropriate to it. Here it realizes itself by bringing about the context in which it joins the being of the thing and surrenders to it, while also recovering itself and illuminating the being which they share with each other: that is, the unitive being which is the thing in its being known by the knower in his. Thought (2), then, has a double ontological relation: to the being of the thing (1) and to the being of its union with the thing (3). These are not simply identical; they are an identification, for they both are and are not the same being. *The relation to the being of the thing* (1): if the copula does not join thought and being, the identification needed for truth is absent. *The relation to the being of their union* (3): but the copula must also distinguish thought and thing in order that there might be knowledge, for only in this distinction can thought return to itself to recover its own being. In so doing it seals itself with its native quality, the transcendental consummation we call *truth.*

The double-nature of the copula as limit is, of course, the source of the ambiguity in its connective function. In affirmative and negative judgments, the copula combines or divides the predicate and the subject: and this is its attributive function. But it also identifies and distinguishes thought and being: and this is its onto-existential function. The degree to which such an identification and distinction becomes explicit will vary from one utterance to another. Nevertheless, since the copula uses some form of the verb, "to be," its onto-existentialist force is at least implicit in all judgments. It is not confined to the controverted judgments of existence ("X is"), nor to judgments which attribute real existence to a subject ("There exists an X such that . . ."). The index of the onto-existential function may be carried by the context within which the judgment is uttered or understood, but it still belongs to the total intention of the judgment. The onto-existential function of the copula, then, is a dense and undeveloped intention to recover the being of things in truth and for thought, an intention launched by thought itself and directing itself to determine the appropriate ontological contexts of the objects of its interest. It belongs to philosophical reflection, however, to render explicit the nature of that function and the interrelation of the modes of being involved.

The attributive and onto-existential functions differ. For the copula *either* combines *or* divides, whereas it must in the very same act *both* identify *and* distinguish. Indeed, even in the negative judgment in which it divides a predicate from a subject, the truthful copula thereby identifies thought with being. It is not enough, however, to simply distinguish the attributive function of the copula, by which it either combines or divides subject and predicate, from its onto-existential weight by which it both identifies and distinguishes thought and being. If we are to recapture the full sense of the judgment, we must also relate these two functions. Since the copula retains its verbal memory, it functions at least implicitly with its ontological heritage of participles and nouns. And so, the combination and division done in its attributive role finds its ultimate justification in the onto-existential function, by which thought identifies the thing as a disclosure of being. But the disclosure is brought about by assertive thought through its affirmations and denials, and so it also discloses the ontological difference between the being of things (1) and the being of thought (2, 3).

It is, of course, possible and even advantageous sometimes to hold in abeyance the ontological power of the copula, and to consider the attributive function in and for itself as a pure connective of terms within a proposition. The non-consideration of the ontological weight of the copula can be carried out in many ways. It can be partial as in traditional logic; or the ontological weight can be formally excluded as in some modern logics. Along such a path of reflection logic can move beyond the simple attributive function to logics based upon other logical relations. Moreover, modal logics are not a straightforward return to the ontological weight of the copula, but an amplification of the modes of attribution. For the "may be" and "must be" of modal logics are considered as possibilities for proposition-formation, and are equivalent to "may be said" and "must be said." Furthermore, other modes of reflective analysis may put some aspect of the ontological weight of the copula out of play, as does the *epoche* which initiates Husserlian phenomenology. So too, in a quite different sense, do the objectives and methods usually employed in the natural and social sciences. In such reductions of the onto-existential force of the copula, thought frees itself to explore structures and processes in and for themselves without having to explicate their relation to their ontological foundations. This dispensation is especially beneficial because the articulation of those foundations in a definite philosophical formulation, such as in the present essay, is not only difficult; since it is a determinate statement, it will inevitably have limitations and defects. Nevertheless, the plethora of possible approaches to the judgment, including its phenomenology, psychology, sociology and logic, must not be allowed to obscure the need for an ontology of judgment. From different perspectives, the present emphasis upon the copula may seem irrelevant, secondary, or even mistaken, but it is an important task of philosophical reflection to revitalize our understanding of the metaphysical and ontological foundations of the judgment. When we put into play the full onto-existential weight of the copula, we commit language and things, thought and being, to a nexus with each other. That association brings about an actual double double-presence within a linguistic and ontological totality.

V

The copula discloses a linguistic totality. The word uttered by a judgment—that something is so and not otherwise—is not the only or the last word. For the copula does not arise out of the mind as a solitary connective, any more than the whole judgment presents itself as an isolated unit. On the contrary, the judgment is an individual and actualized utterance which presupposes a linguistic totality for its ground. It can take form only within such a comprehensive unity. The comprehension is not actual, however, for the totality of actual and possible speech acts is, of course, never present to any speaker. All the possibilities of a language are not at hand at any given moment. Neither is its past wholly recoverable, nor its future wholly predictable. Nevertheless, the "whole" language is "there" in the sense that its forms and functions are available to any skilled speaker. They are not so much *all* there, as they are there in their *allness,* or better, the linguistic possibilities are present in their unity and as a totality which can be drawn upon in determinate ways. If we do not restrict the process of habituation to conscious assimilation or restrict it too closely to acquired habits of particular usage, we might say that language is there as a totality somewhat in the manner of a habit. For it is present with a readiness that is not a fully actual presence, since we are not always speaking, nor is it a mere potentiality for speech, since we are not always learning anew how to speak. Linguists insist that the process of acquiring the structure of a language is a complex one, largely unconscious and appropriated somewhat in the way in which we assimilate other systems and social institutions through enculturation. We come into possession of our native language by a set of processes which bring into play the adoption of motor rhythms and the response to sensory values as well as to meanings. Language, then, is not simply a totality of meaningful sounds. It is an organic adaptation to patterns and possibilities of a distinctive form of behavior, that of the communication of meaning. Never appearing in itself, the linguistic totality is embedded in the speaker's motor and perceptual systems and in his social institutions. Within these possibilities of behavior and by means of assimilated patterns, human consciousness strives for the comprehension and communication of meaning. Language is the availability to the speaker of a totality of communicability in a historically conditioned system of verbal signs and forms of expression. Like a diaphanous medium, this totality is both lit up and shadowed forth through individual acts of speech. Language gathers itself up as a structured totality which breaks forth into concrete and actual expression in individual acts of speech. When the locution is assertive and cognitive, it expresses itself in judgment. When that judgment uses an onto-existential copula, it rises out of a totality of language and thought in order to manifest the quality of truth through a disclosure of being.

VI

The copula discloses an ontological totality, the comprehensive ultimacy of being and non-being. The vigorous verb, "to be" or "not to be," moves easily throughout the language, coupling and uncoupling terms with its discriminatory power of affirmation and denial. Its onto-existential force can, as we have already suggested, be suspended in the interests of exploring factors of language or reality in and for themselves. And it moves so unobtrusively that its onto-existential force and its own intended meaning can be ignored, as when the copula is taken as a mere placeholder which is supposed merely to make room for the union of terms or for preventing their union. It can be all but rendered invisible and ineffective by assigning even its unifying function to the other parts of the judgment. Thus Bosanquet agrees with Mill's view

> that we really need nothing but the subject and predicate, and that the copula is a mere sign of their connection *as* subject and predicate.[5]

The unobtrusiveness which permits the copula to insinuate itself modestly between the terms can easily be mistaken for barren impotence. But the ubiquity, which makes it so obvious that it can be overlooked, arises out of its comprehensive intention. For the copula means to submit the subject and predicate in their relationship to the measure of being and non-being. Indeed, the copula breaks open into its unrestricted universality just because, in breaking open into "is" and "is not," it breaks open into a transcendental value, a value which pervades both speech and things, which counts both in truth and in being. For in asserting the relatedness or non-relatedness of a subject and predicate an ontological measure is elicited through the copula by thought in order to apprehend and appraise the relationship. The disjunct, "is" or "is not," releases the decisive value into which all affirmation and denial is to be ultimately resolved. Within the totality of language, then, the copula discloses itself as the density of all possible affirmations and denials. The totality of assertive speech is gathered together and rooted in the distinction of being and non-being.

Such an onto-existential copula reveals its capacity to be more than a mere place-holder. For the intention of the copula is not only comprehensive, it is also radical and ultimate. With assertive force it carries through a probe which reaches down to the fundamental value associated with truth, to a probity which exceeds the bare coupling and uncoupling of terms, and which authorizes and authenticates the attributive function. As a compo-

---

[5] B. Bosanquet, *The Essentials of Logic*, Macmillan, London, 1895, p. 99, referring to J. S. Mill, *Logic*, Bk. I, c. iv, #1.

nent of linguo-thought, the onto-existential copula does not forget its primitive verbal meaning, nor the familial associations of "is," "exists" and "being." And so it arises out of linguo-thought with the intention to measure and declare what is and is not, was and was not, will and will not, can and cannot, must and must not, ought and ought not.[6] Taken with its ontological weight the copula expresses something which neither the subject alone nor the predicate alone nor both together can say: that their unity or disunity is grounded ultimately neither in an analytic implication nor in a synthetic coincidence of thought and sensible being, but in some appropriate condominium of being and truth. Taken in its assertive sense—that this *is* so and not otherwise—the copula expresses the thrust of the mind towards a true statement which is intended to disclose whatever mode of being is warranted and proper for the declared relationship. Consider these assertions: "The *physical* components of Lear are the markings on a page and the voice and movements of an actor." "He played Lear well *tonight*." "Lear, after all, is a *dramatic character*." What is such speech about? It directs itself towards a grand vague unity which gathers to itself printed page, voice and gesture, a tradition of acting, a history of performances, a context of scholarship, the reflections of critics, a theatrical community and more. This in its unfinished totality is Lear, and is the measure of what can truly be said of it—past, present and future. The copula is the energy of the judgment just because it is suffused with the assertive force of the mind's activity as it strives to associate itself and its concepts with whatever regions, modes and contexts of being are appropriate for them. The copula is the very presence of the mind itself in the act of identifying and distinguishing thought and being. In this activity both being and truth disclose themselves as comprehensive and radical values. For the copula is the achievement of truth just because it is the assertive force of the mind under the discipline of the thing. And it is the disclosure of being just because it reaches the onto-logical terminus

---

[6] Each of these needs further analysis in order to disclose the interplay of being and non-being, presence and absence, temporality and atemporality, reality and ideality that constitutes the fully concrete intentions of the copula. The present essay is only one step in the attempt to enrich the copula through reflection.

intended (*ad rem*) inasmuch as the thing referred to is or is not in just the way it is signified in the judgment.

VII

The copula discloses a double double-presence. Judgment establishes a unity of thought and being, while preserving their distinctness. The meeting of thought and being is, then, a kind of twosome; but what kind? Judgment does not result in two entities, such as a being in knowledge and a being outside knowledge, for that sacrifices their union to a dualism of known and unknowable, against which Berkeley protested. Nor can judgment result in two entities of which one is an original and the other a copy, for this presupposes knowledge rather than explains it, as Sartre and others have shown. For similar reasons I do not think that knowledge consists in the acquisition of a property or state, although human knowing may involve both. The division into "is" and "is not" which is intrinsic to judgment provides a better clue to the character of knowledge, for it arises out of the needs of linguo-thought. The division exhibits a decisive tension within thought itself in the face of being. Hegel takes the division to be self-negation and seeks the reconciliation of thought with itself through the medium of being. The resultant dialectic throws a brilliant light upon many aspects of thought and being. To be sure, thought is not indifferent to its intrinsic division, but the secret power of thought does not lie in its self-division and self-negation alone. It is rooted even more basically in its complicity with being, and in the onto-existential function by which it both identifies and distinguishes itself and being.

Moreover, its complicity with being is rooted not only in the being of the thing, but just as basically in its own being. The complicity shows that thought is not autonomous. From the side of the thing, thought is governed by a capacity which the thing holds within itself: the intelligibility of its being. If thought must serve under the law of the thing, then, without the intelligibility which the thing has in virtue of the intelligibility of its being, thought could not itself be intelligent, that is, could not be thought. But this intelligibility rises up from within thought, too. For, because being is intelligible in the thing, and because thought is itself a manner of being, thought contains a similar capacity within itself: viz., the possibility of disclosing its own intelligibility to itself in virtue of the intelligibility of its being. Such a possibility is the promise of truth. For in realizing the truth of something else, thought vindicates the mutual intelligibility of the being of the thing and the being of thought itself (*ens et verum convertuntur*). This is the first double-presence: the mutual presence of the being of thought and the being of the thing.

Even more fundamental to thought than its intrinsic self-division by which it declares either-or, then, is the doubling of presence by which it establishes both-and. For thought brings together what is in some sense two: knower and known, thought and thing (the first double-presence). It can do this, however, only out of a more original possibility, by which it brings about the doubling of what remains one. When a thing is known, it *is* both in its own being and in its being known. Now being known is not simply being, or everything that is would be known; and yet it is one and the same being that both is and is known, or nothing would ever be known. This second double-presence is the peculiar consummation of the wedding of thought and being. (It is *being* in the third sense noticed above.) Thought opens out onto the thing (*adesse*), even as the thing opens out in the presence of thought (*adesse*). The intelligence of thought realizes itself as the actual intelligibility of being. In doubling what remains one, thought arrives at the truth of something, both distinguishing itself from the thing and identifying itself with it.

The decisive word which being utters in the doubling brought about by judgment is a sort of excess. It is not, however, the excess of a remainder or a redundancy, the superfluity which the early Sartre found absurd and irrational. To call cognition an "excess" may shock some philosophers who secure the respectability of thought by appealing to its economy and necessity, for it seems to make thought unnecessary, useless, and trifling. What "excess" means here, however, is that the possibility of intelligence is rooted in the intelligibility of being, and that the intelligibility of being is a certain propensity for self-transcendence by which a thing both holds itself in its own being while it yields that being to another. Intelligibility is a certain abundance and generosity written into being, its capacity for making a gift of itself to thought. Such ontological gifts are not always easily received, of course, for they do not yield themselves indiscriminately. Yet it is the peculiar capability of thought that it can coax a thing out of itself, at least to some degree. It can do this, undoubtedly, because in being known the thing does not really go out of itself but only enters upon an enlargement of its possibilities. For as it becomes grounded in the being of thought, it remains itself and also remains within a context of being, although that context is a modulation of its being. The possibility of cognition is rooted in this ontological generosity, and in a vicariosity which permits one being to bear the presence and meaning of another.

The copula provides the principle of limit for such a double double-presence. From the side of the thing, and in accordance with the law of the thing, there can be no two entities present, but only the one and same being being known. For the truth arrived at is the truth of *that* being, and not of some other. But there is also duality, because simply being is not actually being known. The presence is double, then, yet a single thing is present. The copula is the limit and focal point of this ontological double-presence. The thing loans its being, but in the loan it does not detach its being from itself. Rather, it offers up its being to a modulation, an enlargement of its horizon, so that it can be in association with another of a peculiar sort, a knower. From the side of thought, this double-presence of the thing known is grounded in the co-presence of thought and being. For in the judgment and through the copula, thought sustains itself and the thing in their union. Indeed, thought makes itself present as receiving the loan which the thing makes of its own being. Thought loans its being, too, then, for it is present to the thing as sustaining it in a relation which lies beyond the power of the thing itself. This community of thought and thing is for the sake of the truth of being. Cognition, then, is brought about by a reciprocity of loaned modes of being, in which thought brings the energy of its active attention and the thing brings the intelligibility of its being. Cognition is a double double-presence, in which both thought and the original being of the thing form a new actuality, the actual community of knower and thing known. The copula functions as the ontological limit for thought and for the being to which the truth must gravitate as to its objective measure: *pondus cognoscentis rei cognoscendae pondus.*

Trinity College,
University of Toronto.

THOMAS AQUINAS ON ANSELM'S ARGUMENT

MATTHEW R. COSGROVE

THOMAS CRITICIZES on five separate occasions the argument for God's existence given by Anselm in the *Proslogion*. The works in which his objections are offered are, in chronological order:[1]

> *In Primum Librum Sententiarum* dist. 3, q. 1, a. 2, 4 & *ad* 4
> *In Boethii De Trinitate* prooem., q. 1, a. 3, 6 & *ad* 6
> *Quaestiones Disputatae De Veritate* q. 10, a. 12, 2 & *ad* 2
> *Summa Contra Gentiles* I, 10 & 11
> *Summa Theologiae* Ia, q. 2, a. 1, 2 & *ad* 2

Of these discussions the last, from the *Summa Theologiae*, is the best known and is often taken as representative of Thomas' response to Anselm.[2] Yet it would seem, on the face of it, unsatisfying as a refutation. Gareth Matthews' comment expresses a very widely shared reaction: "Instead of showing that Anselm's argument is invalid, Aquinas seems content to state, without counterargument, that the alleged conclusion does not follow."[3] To many, Thomas' critique represents no advance beyond Gaunilo in understanding Anselm, but merely reproduces Gaunilo's objection against *Proslogion* III in *Pro Insipiente* VII (namely, that God

---

[1] On the chronology cf. I. T. Eschmann, "A Catalogue of St. Thomas's Works," Appendix to Etienne Gilson, *The Christian Philosophy of St. Thomas Aquinas*, trans. by L. K. Shook (New York: Random House, 1956), pp. 381ff.

Where convenient the titles of these works are abbreviated: *In I Sent.*; *In De Trin.*; *De Veritate*; *CG*; *STh* or *Summa*.

Passages from *CG* and *STh* are quoted according to the standard critical (Leonine) edition of the *Opera Omnia* (Rome, 1882–); quotations from *In De Trin.* follow the text in Bruno Decker, ed., *Expositio Super Librum Boethii De Trinitate*, Studien und Texte zur Geistesgeschichte des Mittelalters, Band IV (Leiden: E. J. Brill, 1959); *In I Sent.* is quoted from the Parma edition of the *Opera Omnia*, vol. 6 (Parma, 1856; repr. New York: Misurgia Publishers, 1948).

[2] It is this passage which is reprinted in Alvin Plantinga's widely used anthology of writings on the proof: *The Ontological Argument. From St. Anselm to Contemporary Philosophers* (Garden City, N.Y.: Doubleday & Co. Inc., Anchor Books, 1965), pp. 28–30.

[3] Gareth B. Matthews, "Aquinas On Saying That God Doesn't Exist," *The Monist*, 47 (1963), p. 474.

must be proven to exist actually before he can be understood as a necessary being),[4] and his critique has even evoked from one of the most influential modern proponents of Anselm's argument the remark "This is not very perceptive, is it?"[5] Such an attitude has, as one might expect, hardly been modified by the satisfaction with which Thomists often regard Thomas' rebuttal while ignoring both the logic of the argument and the details of Thomas' reply to it.[6]

What is most frustrating and perplexing is the apparent and inexplicable failure of Thomas' treatment of the argument to come to terms with what Scotus, Leibniz, and modern writers such as Hartshorne, Malcolm, and Findlay have all regarded as the central issue: the modal character of the argument. If God is "possible," i.e., if the notion of God is not logically contradictory, he exists necessarily. A proof that God does exist in actuality prior to understanding his existence as necessary would be superfluous—this, as the aforementioned writers have all contended,[7] is what Anselm's argument purports to make clear. But this, it would seem, is precisely what Thomas did not see. And yet we would expect him to see it. He was not unacquainted with modal logic, nor with the idea on which the argument turns, for he knew Aristotle's proposition from the *Physics:* "in the case of eternal things, what can be must be."[8] Did he really fail to understand the argument? If so, can this failure be accounted for? If he did understand it, what was in fact his objection to it?

---

[4] See M. J. Charlesworth, *St. Anselm's* Proslogion (Oxford: The Clarendon Press, 1965), p. 88 n. 2, and Arthur C. McGill in *The Many Faced Argument,* ed. by John Hick and A. C. McGill (New York: The Macmillan Company, 1967), p. 88 and n. 187.

[5] Charles Hartshorne, *Anselm's Discovery* (La Salle, Ill.: Open Court, 1965), p. 156.

[6] E.g., Ian Hislop, "St. Thomas and the Ontological Argument," *Contemplations* (Oxford: Blackfriars, 1949), pp. 32–38; Etienne Gilson, *St. Thomas Aquinas,* pp. 54, 56. See also n. 32, below.

[7] Among them, J. N. Findlay concludes on the negative side of the question, holding that God is impossible or that the notion of God is meaningless. See his "Can God's Existence Be Disproved?" *Mind,* 57 (1948), pp. 176–83; repr. in Plantinga, pp. 111–22, and elsewhere.

[8] *Physics* III.4, 203b30. Discussing this passage of the *Physics,* Thomas paraphrases the above proposition, in Book III Lecture 7 of his commentary *In VIII Libros Physicorum Aristotelis,* but he does not go beyond the application of it to the problem at hand there, i.e., the existence of an infinite body. No mention is made of its possible theological import.

It is not difficult to understand why *in general* Thomas opposed the argument. His broad philosophical concerns leave no doubt that the tendency represented by the argument would be unacceptable to him. This would be clear even if he had not dealt with the argument at all. The program with which the *Summa Theologiae* begins—i.e., the justification of theology as a science and a wisdom in the Aristotelian sense of these terms—implies a philosophical situation in which the relation between faith and reason has become critical for both philosophy and theology in a way that it certainly was not yet for Anselm. Furthermore, the *Proslogion*, by virtue of its personal and confessional character (contrast its first chapter with the first question of the *Summa*) and its search for truth through the inner man, openly adheres to the Augustinian tradition's emphasis on the certainty of what is interior, in contrast to the uncertain things of sense, as the basis of knowledge.[9]

At the same time, however, the *Proslogion*'s search for *unum solum argumentum,* as well as the highly developed logical concern revealed by the form of the argument and, indeed, Anselm's entire philosophical activity,[10] show Anselm significantly departing from that tradition[11] to the extent that he was attempting to give theology a new basis by establishing its first principle as necessary and certain for reason.[12] With this theological goal Thomas was in sympathy, and he takes Anselm seriously enough to recognize that his argument is of great importance for the foundations of theology and must be dealt with in beginning his own attempt to make theology *scientia*.[13]

But by the time of Thomas the argument, in the hands of its philosophically Augustinian proponents, had become a tool to op-

---

[9] Cf. Augustine *De Libero Arbitrio* II.8.21.

[10] Cf. Desmond P. Henry, *The Logic of St. Anselm* (Oxford: The Clarendon Press, 1967).

[11] Anselm's considerable efforts to develop theology in a deliberately more philosophical direction than the traditional Augustinianism then current are discussed by H. Liebeschütz in his chapter on Anselm in *The Cambridge History of Later Greek and Early Medieval Philosophy*, ed. by A. H. Armstrong (Cambridge: Cambridge University Press, 1967), pp. 611–23, *passim*.

[12] Cf. Heribert Boeder, "Die Fünf Wege und das Princip der thomasischen Theologie," *Philosophisches Jahrbuch*, 77 (1970), p. 77.

[13] *Ibid.*, pp. 77–79.

pose such a theology. This is clearest in the case of Bonaventure. A Franciscan (which is to say an Augustinian), he was Thomas' contemporary (they were admitted together to the University of Paris by Papal intervention in 1257) and philosophical antipode. He knew Anselm's argument and he accepted it.[14] But for Bonaventure the argument had become "God is God, therefore he exists," with which—as Etienne Gilson remarks—"the dialectical process is now simplified . . . to the point of vanishing altogether."[15] It had become a means for the reduction of theology's foundation to tautology and the condemnation of Thomas' entire program.[16]

It is not therefore entirely an exaggeration to see the core of a complex of philosophical and theological issues in the question which Anselm's argument, primarily via its adaptation by Bonaventure, had become for Thomas, i.e., in the question: *Utrum Deum esse sit per se notum,* "Whether 'God exists' is self-evident" (*STh* Ia, q. 2, a. 1). Those who object that Anselm himself did not hold God's existence to be self-evident and that in dealing with Anselm under this heading Thomas completely misconstrues the argument of the *Proslogion* have, to begin with, ignored the significance with which the immediate historical context had endowed it. But there is in addition some question as to whether Thomas' well known use of the Aristotelian distinction between the two kinds of self-evidence is in fact the essence of his critique of Anselm's argument, as has been supposed,[17] a question that may be considered

---

[14] See Bonaventure *In I Sent.* dist. 8, p. 1, a. 1, q. 2; *De Mysterio Trinitatis* q. 1, a. 1, 21-24; *In Hexaemeron* coll. V, 31.

[15] Etienne Gilson, *The Philosophy of St. Bonaventure,* trans. by Dom Illtyd Trethowan and Frank J. Sheed (New York and London: Sheed & Ward, 1938), p. 116.

[16] The fundamental opposition of Bonaventure and Thomas in this matter, the starting point of theology, is not affected by the dispute between Gilson and Van Steenberghen over Bonaventure's philosophical relation to Aristotelianism. For a summary of that dispute and defense of Gilson's view of Bonaventure as Augustinian in philosophy, and not only in theology, see David Knowles, *The Evolution of Medieval Thought* (Baltimore: Helicon Press, 1962), pp. 243-46.

[17] E.g., by Gareth Matthews, pp. 472-73, and by Karl Barth, *Anselm: Fides Quaerens Intellectum,* trans. by Ian W. Robertson (Cleveland and New York: World Publishing Co., Meridian Books, 1962), p. 78 n. 2.

once his discussion of it in the *Summa Theologiae* has been examined in detail. Is Thomas' rebuttal of the *Proslogion* as found in the *Summa* as mistaken, or irrelevant, or as "crude" [18] as it has appeared? Is it no more than "a dogmatic denial that the proof is valid?" [19]

Of the five passages listed earlier, the *Summa Theologiae* provides the most appropriate focus for examining Thomas' critique of Anselm. It is not only the latest chronologically and the best known; but, considering the importance within the *Summa* of the question *Utrum Deum esse sit per se notum* in comparison with Thomas' seemingly inadequate response to Anselm's position, it is also the most philosophically perplexing statement of that critique. It provides such a focus moreover because each of Thomas' discussions of Anselm may best be dealt with in relation to it.

It will be convenient at this point to summarize the objections, mentioned at the outset, which have been brought against those discussions. There are essentially two objections: 1) Anselm did not hold that God's existence is "self-evident" and Thomas is wrong to deal with his argument under this heading. The formula rejected by Thomas is consequently not Anselm's formula and "his criticisms rest upon a misunderstanding." [20] 2) Thomas fails to grasp the modal character of Anselm's argument and nowhere offers any rebuttal which even attempts to come to terms with this feature of it, much less one which would effectively counter it. [21] The presentation which follows is directed against these objections.

In the *Summa Theologiae* Thomas gives Anselm's reasoning thus:

> ... those things are said to be self-evident which are known as soon as their terms are known. ... But once it is understood what this name *God* signifies, it is immediately grasped that God exists. For by this name is signified that than which a greater cannot be signified. But what exists in reality and in the intellect is greater than what exists only in the intellect; whence, since once this name is understood God is immediately in the intellect, it also follows that

---

[18] Hartshorne, *Anselm's Discovery*, p. 160.
[19] *Ibid.*
[20] Charlesworth, pp. 5, 58–59, and 58 n. 1; cf. also n. 19, above.
[21] Hartshorne, *Anselm's Discovery*, pp. 156 and 160ff.

he exists in reality. Therefore [the proposition] "God exists" is self-evident.[22]

Two aspects of this passage deserve comment before proceeding. First, it will be noticed that for Anselm's word *"cogitari"* in the formula *id quo maius cogitari non potest*, "that than which a greater cannot be thought" (*Proslogion* II), Thomas has substituted the word *"significari."* However, no important change is thereby effected; elsewhere Thomas does use the word *"cogitari"* in giving Anselm's formula,[23] and clearly considers *"significari"* and *"cogitari"* equivalent, for the arguments he brings against either formulation are the same. Secondly, in partial answer to the first of the two objections cited above, it is to be observed that while Thomas does take Anselm to hold that "God exists" is *per se notum,* he is nevertheless aware that Anselm *argues* for that position, and he presents the reasoning of *Proslogion* II correctly.[24] Hence it is somewhat beside the point to object, as Charlesworth does, that "it is principally because the *Proslogion* argument is formally and logically an *argument* that the proposition 'God exists' is not for Anselm *'per se notum'* or analytic or self-evident."[25] What is relevant is not whether Thomas associates the argument with a derivative position without projecting Anselm's possible attitude toward such an association, but solely whether Thomas does justice to the reasoning of the argument itself.[26]

---

[22] ... *illa dicuntur esse per se nota, quae statim, cognitis terminis, cognoscuntur. . . . Sed intellecto quid significet hoc nomen Deus, statim habetur quod Deus est. Significatur enim hoc nomine id quo maius significari non potest: maius autem est quod est in re et intellectu, quam quod est in intellectu tantum: unde cum, intellecto hoc nomine Deus, statim sit in intellectu, sequitur etiam quod sit in re. Ergo Deum esse est per se notum.* (Ia, q. 2, a. 1, 2.)

[23] Cf. *In I Sent.* dist. 3, q. 1, a. 2 and *CG* I.10.

[24] Karl Barth and those sympathetic to his fideistic interpretation of Anselm will object that no presentation of the argument which ignores its context within the prayer introduced by *Proslogion* I can be called "correct." It is not possible to consider this position here, but it may be noted that such a context would at least no longer have been relevant to the philosophical status of the argument faced by Thomas.

[25] Charlesworth, p. 58 n. 1.

[26] But cf. n. 30, below, on the question as to whether Anselm's own position does in fact assert such self-evidence.

We meet the occasion of that above objection in the *Responsio* to the problem of the article, *Utrum Deum esse sit per se notum*. After discussing the two modes of self-evidence, distinguishing between a proposition which is self-evident in itself but not to us (*secundum se et non quoad nos*) and one self-evident both in itself and to us (*secundum se et quoad nos*), Thomas continues:

> I say therefore that this proposition "God exists," taken in itself, is self-evident, because the predicate is the same as the subject; for God is his existence, as will be shown below. But, since we do not know concerning God what he is, [that proposition] is not self-evident to us, but needs to be demonstrated through those things which are better known to us and less known according to nature, namely through [his] effects.[27]

Is this Thomas' answer to Anselm's argument? It might seem so, particularly since in *In De Trinitate* this distinction is the only rebuttal to Anselm which he advances. There, however, it is not Anselm's argument, but only the result of it, from *Proslogion* III, which is adduced in support of the question at hand (*Utrum Deus sit primum quod a mente cognoscitur*, "Whether God is the first thing which is known by the mind"):

> ... nor is it possible for God to be thought not to exist, as Anselm says.[28]

To which Thomas replies:

> [the proposition] "God exists," taken in itself, is self-evident, because his essence is his existence, and in this manner speaks Anselm. Not, however, [self-evident] to us, who do not see his essence. Nevertheless, the knowledge of him is said to be innate in us, insofar as by means of principles innate in us we are easily able to perceive that God exists.[29]

---

[27] *Dico ergo quod haec propositio*, Deus est, *quantum in se est, per se nota est: quia praedicatum est idem cum subiecto; Deus enim est suum esse, ut infra patebit. Sed quia nos non scimus de Deo quid est, non est nobis per se nota: sed indiget demonstrari per ea quae sunt magis nota quoad nos, et minus nota secundum naturam, scilicet per effectus.*

[28] ... *nec potest Deus cogitari non esse, ut dicit Anselmus.* (*In De Trin.* prooem., q. 1, a. 3, 6.)

[29] ... *Deum esse, quantum est in se, est per se notum, quia sua essentia est suum esse—et hoc modo loquitur Anselmus—non autem nobis qui eius essentiam non videmus. Sed tamen eius cognitio nobis innata esse dicitur, in quantum per principia nobis innata de facili percipere possumus Deum esse* (*Ibid.*, ad 6.). Decker ed., pp. 70 (line 7) and 73 (lines 30–32)–74 (lines 1–2).

What is the function of that distinction, there and in the *Summa?* It is not an answer to Anselm's argument, and this is quite clear in the context of the *Summa,* for in *ad* 2 Thomas will go on to deal with Anselm in particular. In the *Responsio* his purpose is somewhat different. It is to explain why, if God's existence is self-evident in the sense in which it *is* affirmed to be by Thomas (in the sense that God's essence is his existence, that his existence is *per se notum secundum se*), it is yet capable of denial. This is possible, he answers, because the proposition "God exists," while self-evident *secundum se,* is *not* self-evident *quoad nos.* It is not known immediately, but is the subject of demonstration by sacred doctrine, which demonstration is necessary for us, if belief in God's existence is to be knowledge.

It is not Anselm's argument, but his conclusion, to which the distinction between two modes of self-evidence is directed. That is to say, it is directed against the belief that God's existence cannot be denied, that God cannot be thought not to exist, save—as Anselm would have it—by someone who is simply *stultus et insipiens* (*Proslogion* III).

Thomas recognizes that the non-believer, too, has the possibility of appealing to reason; he can deny God's existence, as Boeder puts it, without contradiction, without giving up the basis of the sciences.[30] So far from being *stultus et insipiens,* in support of his unbelief he can bring forth arguments, as Thomas does in his name in article 3 of the *Summa's* second question, and what answers his denial is not reason outraged by an absurdity but God's revelation of himself: "I am who am" (*Sed Contra* of that article). The rationality and scientific character of theology and of its knowledge of the principle, "God exists," does not reside in that principle's self-evidence or undeniability *quoad nos.*[31]

But it is clear that the foregoing presupposes, and is not itself, Thomas' rebuttal of Anselm's argument. For it was the point of that argument to show that God's existence was such that he could not be thought not to exist, but could only be denied by

---

[30] Boeder, p. 79.
[31] *Ibid.,* pp. 74 and 79. It resides in the end, as Boeder has shown, on that principle's self-evidence *secundum se,* i.e., on God's knowledge of himself.

a fool.[32] To this Thomas replies in the *Summa* as follows:

> ... perhaps he who hears this name *God* does not understand that there is signified [by it] something than which a greater cannot be thought, since certain people have believed that God is a body. Granted, however, that someone understands that by this name *God* is signified this which is said, namely, that than which a greater cannot be thought, nevertheless it does not follow from this that he understands what is signified by the name to exist in reality, but only in the apprehension of the intellect. Nor can it be argued that it exists in reality unless it were granted that there exists in reality something than which a greater cannot be thought: which is not granted by those who posit that God does not exist.[33]

How can Thomas hope to rebut Anselm's argument with this? For the argument proceeds precisely by showing that once *id quo maius cogitari non potest* exists *in intellectu*, it cannot be understood to exist *in apprehensione intellectus tantum*, or as Anselm puts it *in solo intellectu*. Otherwise, *potest cogitari esse et in re, quod maius est* (*Proslogion* II). Anselm continues in conclusion:

> If therefore that than which a greater cannot be thought exists in the intellect alone, that very thing than which a greater cannot be thought is that than which a greater can be thought. But certainly this cannot be. Therefore there exists beyond doubt something than which a greater cannot be thought, both in the intellect and in reality.[34]

---

[32] Given this, it might be contended that Anselm did hold God's existence to be self-evident in some sense, but that is beside the point here. Rather, conceding the popular objection that Anselm's argument does not amount to the assertion *Deum esse est per se notum*, the purpose here is (in part) to show that Thomas' discussion of self-evidence is still relevant to a critique of the argument.

Nevertheless, cf. Thomas' different definition of self-evidence, with explicit reference to Anselm, in *In I Sent*. dist. 3, q. 1, a. 2, 4: *illud est per se notum quod non potest cogitari non esse*.

[33] ... *forte ille qui audit hoc nomen* Deus, *non intelligit significari aliquid quo maius cogitari non possit, cum quidam crediderint Deum esse corpus. Dato autem* quod quilibet intelligat hoc nomine* Deus *significari hoc quod dicitur, scilicet illud quo maius cogitari non potest; non tamen propter hoc sequitur quod intelligat id quod significatur per nomen, esse in rerum natura; sed in apprehensione intellectus tantum. Nec potest argui quod sit in re, nisi daretur quod sit in re aliquid quo maius cogitari non potest: quod non est datum a ponentibus Deum non esse.* (Ia, q. 2, a. 1, ad 2.)

[34] *Si ergo id quo maius cogitari non potest, est in solo intellectu: id ipsum quo maius cogitari non potest, est quo maius cogitari potest. Sed certe hoc esse non potest. Existit ergo procul dubio aliquid quo maius cogitari non valet, et in intellectu et in re.*

---

* Retaining the reading of the manuscripts, *autem;* the Leonine edition reads *etiam*.

Thomas' reply to the argument does seem, as so many have thought, blithely to ignore it. Why does he insist that it must first be granted *quod sit in re aliquid quo maius cogitari non potest?* [35]

Matthews has suggested that the apparent irrelevancy of this reply is what has led commentators to settle "on his distinction between the two kinds of self-evidence as his most important criticism of Anselm's argument." [36] But as we have already seen, that distinction has a different function, and one which we shall explore still further, below.

Another suggestion has been that Thomas here is considering only the non-modal form of the argument, and that his criticism is relevant only to that.[37] This is the form of the argument as abstracted solely from *Proslogion* II, i.e., without the clarification of it provided by *Proslogion* III, by which we are to understand *"maius"* to imply non-contingency. Anselm's reply to Gaunilo makes it plain that he intended the original argument in this, the modal form, and that he was well aware that only in this form would it be effective (*Reply* I–III): only in the case of necessary being can meaningfulness or conceivability be the basis of an inference as to existence. Thus, taken by itself, *Proslogion* II *would* be subject to the objection that *id quo maius cogitari non potest* need only be understood to exist *in apprehensione intellectus tantum*, and not *in re*, as Gaunilo in fact argues (*Pro Insipiente* II) and as Thomas appears to do above.

Did Thomas have only *Proslogion* II in mind? Is his rebuttal in the *Summa* directed only against this form of the argument? Does he ignore its stronger form?

To all these questions the answer is no. In the commentary *In I Sent.* Thomas gives a succinct presentation of the modal form of the argument, after explicitly ascribing it to Anselm:

> God is that than which a greater cannot be thought. But that which cannot be thought not to exist is greater than that which can be

---

[35] This is the difficulty which Thomists commonly overlook; see n. 7, above.
[36] Matthews, p. 474.
[37] Hartshorne, *Anselm's Discovery*, p. 160.

thought not to exist. Therefore God cannot be thought not to exist, since he is that than which nothing greater can be thought.[38]

Thomas' reply to this is similar to that in the *Summa:*

> ... the reasoning of Anselm is to be understood thus. After we understand God, it is not possible that it be understood that God exists, and that he could be thought not to exist; but from this it does not follow that someone could not deny or think that God does not exist; for he can think that nothing of this sort exists than which a greater cannot be thought; and therefore his [Anselm's] reasoning proceeds from this supposition, that it should be supposed that something does exist than which a greater cannot be thought.[39]

Before considering this rebuttal, let us first re-emphasize that, as the presentation which precedes the above has shown, Thomas was not unaware of the modal character of the argument. It is presented again in the same way in his *De Veritate* (q. 10, a. 12, 2). But there the only reply to it is a discussion similar to that in the commentary on Boethius' *De Trinitate,* quoted earlier, of the two modes of self-evidence. In other words, what is answered there is not the argument itself but only its conclusion. We cannot look there for the treatment of the modal form of Anselm's proof which we are seeking from Thomas.

Does the reply from the commentary on the *Sentences* provide it, or do we face there the same perplexity encountered in the case of the *Summa Theologiae?* Is that reply the same as the *Summa*'s? Does Thomas answer the modal form of the argument no differently than its non-modal form?

There is one difference. Whereas the *Summa*'s rebuttal argues that it is possible to hold that *id quo maius cogitari non potest* exists in the intellect alone, that of the commentary on the *Sentences* might be taken to go a step further. Thomas says "[the

---

[38] *Deus est quo maius cogitari non potest. Sed illud quod non potest cogitari non esse, est maius eo quod potest cogitari non esse. Ergo Deus non potest cogitari non esse, cum sit illud quo nihil maius cogitari potest* (dist. 3, q. 1, a. 2, 4).

[39] ... *ratio Anselmi ita intelligenda est. Postquam intelligimus Deum, non potest intelligi quod sit Deus, et possit cogitari non esse; sed tamen ex hoc non sequitur quod aliquis non possit negare vel cogitare, Deum non esse; potest enim cogitare nihil huiusmodi esse quo maius cogitari non possit; et ideo ratio sua procedit ex hac suppositione, quod supponatur aliquid esse quo maius cogitari non potest* (Ibid., ad 4).

objector] can think that nothing of this sort exists than which a greater cannot be thought." Does he mean "exists neither in the intellect nor in reality"? Or only "he can think that nothing of this sort exists in reality, but only in the intellect"? The latter meaning would, it seems, be equivalent to the reply of the *Summa;* the former would not. The former would be such an answer to the modal form of the argument as Scotus, Leibniz, and modern writers like Hartshorne, Malcolm, and Findlay have demanded: it would be saying that the formula *id quo maius cogitari non potest* is meaningless, i.e., that the idea of God (at least as *id quo maius cogitari non potest*) is incoherent. Is there any reason to suppose that this is what Thomas actually had in mind? Would this mean that Thomas recognized the possibility of, and felt the argument could not be effective against, what Hartshorne calls the "positivist" alternative to theism (in contrast to what he refers to as the "atheistic" alternative, the denial of God's factual existence construed as a contingent rather than as a necessary question)—an alternative which Hartshorne, along with the other writers mentioned, regards as the only viable rebuttal to Anselm's argument?[40]

One more discussion of Anselm in Thomas' work remains to be considered. In the *Summa Contra Gentiles,* after stating (in I.10) the form of the argument given in the *Summa Theologiae* (i.e., that drawn from *Proslogion* II), Thomas replies:

> . . . [although it be] granted that by this name *God* there is understood by everyone something than which a greater cannot be thought, it will not be necessary that there exist in reality something than which a greater cannot be thought. For it is necessary that the thing and the definition of its name be posited in the same way. From this fact, however, that what is proposed by this name *God* is conceived by the mind, it does not follow that God exists, save only in the intellect. Whence it will not be necessary either that that than which a greater cannot be thought exist save only in the intellect. And from this it does not follow that there exists something in reality than which a greater cannot be thought. And so no difficulty befalls those who posit that God does not exist. For it is not a difficulty that given anything either in reality or in the intellect something greater can be thought, save only for him who concedes that there exists something in reality than which a greater cannot be thought.[41]

---

[40] Hartshorne, *Anselm's Discovery,* pp. 53ff.
[41] . . . *dato quod ab omnibus per hoc nomen* Deus *intelligatur aliquid quo maius cogitari non possit, non necesse erit aliquid esse quo maius*

The final sentence is the key: Anselm's argument presents no difficulty for one who holds that for anything given in reality *or* in the intellect, a greater can be thought. This, as Gareth Matthews has pointed out in his illuminating analysis of the passage,[42] is the logical equivalent of the proposition "There is nothing than which a greater cannot be thought," and this is an effective reply to either the non-modal or the modal form of the argument, for only the proposition "There is something than which a greater cannot be thought, but it exists in the intellect alone, and not in reality," is countered by the argument. Here we have a clear statement of what previously, in the case of his reply from the commentary on the *Sentences,* we were only able to conjecture *might* have been Thomas' meaning. That is to say, here Thomas clearly is suggesting that it is possible to oppose Anselm by holding that *id quo maius cogitari non potest* is an impossible or meaningless conception, since a greater can always be thought.

There is a difficulty in this passage which Matthews does not discuss. Why does Thomas say that for one holding this position, *id quo maius cogitari non potest* need not exist *save only in the mind (nisi in intellectu)*? For, as we have seen, in order to resist the argument it must be denied that this notion exists *even* in the mind. We must therefore suppose that here (and elsewhere) when Thomas says that it may be held to exist only mentally and not actually, he means that it is held to exist mentally in the sense that it is entertained and rejected by thought, in terms both of actual and conceptual existence strictly understood. This inference as to his meaning not only seems justified in the light of Thomas' demonstrated grasp of the argument's modal concepts, but also is required by the conclusion of the preceding passage

---

cogitari non potest in rerum natura. Eodem enim modo necesse est poni rem, et nominis rationem. Ex hoc autem quod mente concipitur quod profertur hoc nomine Deus, *non sequitur Deum esse nisi in intellectu. Unde nec oportebit id quo maius cogitari non potest esse nisi in intellectu. Et ex hoc non sequitur quod sit aliquid in rerum natura quo maius cogitari non possit. Et sic nihil inconveniens accidit ponentibus Deum non esse: non enim inconveniens est quolibet dato vel in re vel in intellectu aliquid maius cogitari posse, nisi ei qui concedit esse aliquod quo maius cogitari non possit in rerum natura.* (*CG* I.11.)

[42] Matthews, p. 475.

with its suggestion that one might hold that something greater can be thought than anything given either in reality or in the intellect.

Hartshorne has raised an objection to Matthews' examination of that passage with which we must briefly deal. The proposition "There is nothing than which a greater cannot be conceived" may, he argues, be offered as either a contingent or a necessary truth. But if offered as a contingent truth it is contradictory:

> If contingent, there must be no logical impossibility in the existence of a not conceivably surpassable being. But since, according to [that proposition], there is in fact no such being, its non-existence is also taken as possible. It would follow that the not impossible existence of an unsurpassable being could only be contingent existence. But . . . a contingent being could not be unsurpassable. Thus [taken as a contingent truth the proposition] is contradictory.

It must therefore, Hartshorne continues, be offered as a necessary truth if it is to stand up to Anselm's argument:

> A necessity that, given any being, a greater can be conceived implies the logical impossibility of an unsurpassable being. This, however, is the positivistic not the atheistic tenet. Moreover, if a concept is logically impossible, this can be no mere truth of fact. Modal statements themselves, as Aristotle saw, have the mode of necessity, not of contingency. We conclude that atheism (the merely factual denial of God's existence) is not saved from contradiction by Thomas. Neither the divine existence nor the divine nonexistence could be a mere fact, i.e., a contingent truth. The question is conceptual not observational. Anselm correctly located the theistic issue in the logical landscape. Did Aquinas? If indeed the tenability (or at least initial plausibility) of positivism was his objection, this never becomes very clear in his discussion. (And Gaunilo had already made the point quite as definitely.)[43]

But had Gaunilo actually made that point? Hartshorne is presumably referring to the objection advanced in *Pro Insipiente* that we do not, strictly speaking, have an idea of God in the mind. We do not, Gaunilo had argued, because while even false and nonexistent things can be entertained in the mind by virtue of their posited resemblance to things that are familiar, there is no basis on which the notion "God" or "that which is greater than all things" (*maius omnibus*) could be so posited. He who hears these words thinks them not as an idea in his mind but as an

---

[43] Hartshorne, *Anselm's Discovery*, pp. 161–62.

affection of the mind (*secundum animi motum*), while trying to imagine what they might mean (*Pro Insipiente* IV). Hence, he continues, it cannot be proven that this *maius omnibus* exists in reality, since it is not admitted that it exists even in the mind (*Pro Insipiente* V). Is this the same point as Thomas is making?

Thomas hews more closely to the argument. The objection Gaunilo had made only seems to resemble Thomas'; there is a crucial difference. Gaunilo's objection posed no difficulty for Anselm because, briefly put, Gaunilo wrongly supposed that "*id quo maius cogitari non potest*" and "*maius omnibus*" are equivalent. But, as Anselm replied, these formulae are not equivalent, and the difficulty that can be raised against the possibility of conceiving what is meant by "*maius omnibus*," and conceiving it in such a way that it might be said to exist "in the mind"—that difficulty does not apply to the possibility of conceiving *id quo maius cogitari non potest* (*Reply* V). It had not occurred to Gaunilo to hold, what Thomas sees an opponent of the argument might do, that there is *for thought* nothing than which a greater cannot be thought.

Nor does Thomas' objection require, as Hartshorne maintains it does, that Thomas had in mind "the tenability (or at least initial plausibility) of positivism." It is not the impossibility of God existing, but the impossibility of conceiving something so great that a greater cannot be conceived, which Thomas thinks a plausible position. And this, *pace* Hartshorne, does not imply "the logical impossibility of an unsurpassable being"; it implies only the impossibility of conceiving that unsurpassability. And it implies further the unlikelihood of the possibility (if not the impossibility simply) of conceiving it as an hypothesis merely, rather than as something which would only be sought to be comprehended by a mind convinced of its existence by other reasons. Thus to hold that there is nothing than which a greater cannot be conceived is not to hold that God *necessarily* does not exist.

Thomas is clearly not concerned to argue against the "positivistic" position when he goes on to prove by his "five ways" that God exists, and he is not so concerned because in treating Anselm's argument he has not raised such a position and has not had to do so in order to rebut Anselm. Opposition to the argument does not require a "proof" of God's impossibility; an assertion about what can and cannot be *thought* by man, not an assertion about what can and cannot exist *secundum se*, suffices. And this is the primary reason why, in his discussions of Anselm, the distinction between the two modes of self-evidence is so important for Thomas. Given the resistance to Anselm's argument which he regards as possible to maintain, Thomas must explain *why* it is possible. That is, he must account for how what is supremely self-evident *secundum se* can yet be claimed inconceivable. He must account for why someone could not only think what, properly speaking, cannot be thought (God's non-existence), but also—what is more immediately relevant—why someone can fail to see that *he himself can* think something than which he can think nothing greater. This is only possible, in short, because—as Thomas says, quoting Aristotle—"our intellect is related to those things which are most knowable as the eye of an owl is to the sun" (*CG* I.11).[44]

Thomas' use of the distinction between the two modes of self-evidence in dealing with Anselm's argument lays the basis for overcoming the position whose power against the argument he had recognized and for proceeding with his own proof of God's existence. This will not involve a logical demonstration of the meaningfulness or conceptual possibility of the idea of God; even Scotus, who later was to "color" Anselm's argument by stipulating that the concept of God first be shown to be the concept of a possible being,[45] saw that this could not be shown a priori, but only by recourse to a "cosmological" argument.[46] Rather, it is to be overcome by proceeding through the distinction itself, from what is evident to us (but not in itself) to what is evident in itself and to us. This is appropriate, since the basis of the possibility of opposing Anselm's argument is that God's existence is evident in itself but not to us, since that opposition is possible because of that distinction. Thomas' own proofs, the "five ways," by proceeding

---

[44] Aristotle *Metaphysics* II.1 993b9. Actually, Aristotle speaks of a bat.

[45] *Ordinatio* I, dist. 2, p. 1, q. 1–2, 137–38 (*Ioannis Duns Scoti Opera Omnia* [Vatican City: Vatican Press, 1950], vol. 2, pp. 209–10).

[46] Cf. Frederick Scott, "Scotus, Malcolm, and Anselm," *The Monist*, 49 (1965), pp. 634–38.

according to that distinction, are arguments against Anselm's opponent, who by virtue of the distinction (or more precisely, the state of affairs to which the distinction applies) can hold (a) that there is nothing than which a greater cannot be thought, consequently (b) that God can be thought not to exist, and (c) that God does not exist.

That Thomas does seek to overcome Anselm's opponent is obvious in the first instance because, in denying that God exists, Anselm's opponent is of course Thomas' opponent as well. But there is another aspect of Thomas' attempt to deal with him which bears a closer relation to Anselm, one which is also more specific than the sympathy of general theological aims discussed earlier. This particular, positive relation to Anselm does not, however, consist for Thomas—as it did for Scotus—in a desire to rehabilitate the argument and make of it, if not a proof, a *persuasio probabilis*. It consists rather in the fact that his goal, namely to make what is self-evident *secundum se* evident *quoad nos,* will also show that it *is* possible to conceive *id quo maius cogitari non potest,* and will do this precisely in the way in which Anselm had seen that it must be done: through contingent things (cf. *Reply* VIII). It is consequently a mistake to suppose that Anselm's *Proslogion* argument dispenses with the need for a cosmological argument,[47] or that a return to it might "open a new era in metaphysics."[48]

It has become clear, then, that Thomas' emphasis on the distinction between modes of self-evidence in his treatments of Anselm's argument is by no means irrelevant. Though this distinction is not itself his criticism of the argument, as has often been thought, it is nonetheless of far-reaching importance for his true objection and for the ultimate outcome of that objection.

Finally, it should by now also be clear that Thomas cannot be criticized[49] for a failure to distinguish between the "two forms"

---

[47] Hartshorne, *Anselm's Discovery,* pp. 157 and 234, and Aimé Forest in *Le Mouvement doctrinal du IXe au XIVe siècle* (Paris: Bloud and Gay, 1951), p. 61 (cited by Charlesworth, p. 61 n. 2).

[48] Charles Hartshorne, *Creative Synthesis and Philosophic Method* (La Salle, Ill.: Open Court, 1970), p. 55.

[49] Hartshorne, *Anselm's Discovery,* pp. 155ff.

of the argument—the non-modal form drawn from *Proslogion* II alone (the form attacked by Gaunilo with the famous "Lost Island" example in *Pro Insipiente* VI; a form never intended by Anselm), and the modal form as in *Proslogion* II and III taken together, or as given in the *Reply*. We have seen (1) that Thomas realized how *Proslogion* II and III function together as one argument (most plainly in *In I Sent.* and *De Veritate*); (2) that his objection does deal with the modal character of the argument (cf. *In I Sent.* and *CG*); (3) that where he separates *Proslogion* II and *Proslogion* III (as in *CG*) or provides a summary of only one of these chapters (*Proslogion* II in *STh; Proslogion* III in *In De Trin.*) he is dealing separately with the argument's premise (something can be thought than which a greater cannot be thought—as in *STh* and *CG* I.10.2) or its conclusion (God cannot be thought not to exist—as in *In De Trin.* and *CG* I.10.3) in a way that can be understood as valid in terms of that objection. In some of these passages one could wish for more explicit discussion. But one could not ask of Thomas a better understanding of Anselm's argument.

The Catholic University of America.

SAINT THOMAS AND SIGER OF BRABANT REVISITED
EDWARD P. MAHONEY

OF ALL THE THINKERS who were contemporaries of Saint Thomas, perhaps none has so intrigued historians of medieval philosophy and evoked such scholarly controversy as has Siger of Brabant (*b. ca.* 1240; *d.* 1281–1284).[1] Given the fact that Dante places Siger along with such illustrious minds as Saint Albert the Great, Dionysius the Areopagite, and Saint Thomas himself in the fourth heaven described in the *Divine Comedy,* it is not surprising that scholars attempted to find a strong influence of Thomas on Siger, especially in the great controversy regarding the unity of the intellect—some even appeared to suggest he had become a Thomist.[2] Indeed, by reason of the ascription to Siger of questions on the *De anima* found in manuscripts in Munich and Merton College, Oxford, which are strongly Thomistic in orientation, it was argued that Siger had eventually drawn close to the Thomist position in psychology.[3] Although most Siger experts now doubt the ascription of those questions to Siger, or at least those on Book III,[4]

---

[1] For a brief account of Siger's life, see P. Glorieux, "Siger de Brabant," in *Dictionnaire de théologie catholique,* XIV (Paris, 1941), cols. 2041–2042. For good reviews of scholarship on Siger, see Van Steenberghen, "Nouvelles recherches sur Siger de Brabant et son école," *Revue philosophique de Louvain,* LIV (1956), pp. 130–47; Armand A. Maurer, "The State of Historical Research in Siger of Brabant," *Speculum,* XXXI (1956), pp. 49–56; Albert Zimmermann, "Dante hatte doch Recht; Neue Ergebnisse der Forschung über Siger von Brabant," *Philosophisches Jahrbuch,* LXXV (1967/68), pp. 206–17.

[2] The passage in Dante was much debated by Busnelli, Perugini, Nardi and other Italian scholars. See Etienne Gilson, *Dante and Philosophy,* trans. D. Moore (New York, 1963).

[3] See F. Van Steenberghen, *Siger de Brabant d'après ses oeuvres inédites,* 2 vols. (Louvain, 1931 and 1942). The ascription of these questions to Siger was vigorously contested by Bruno Nardi in many of his publications. See *Giornale critico della filosofia italiana,* XVII (1936), pp. 26–35; XVIII (1937), pp. 160–64; and XX (1939), pp. 453–71, for some of his critiques of Van Steenberghen. Gilson (pp. 317–27) supported Nardi.

[4] It would be impossible to review the scholarship on the question here. For good introductions, see Maurer, pp. 50–53; Zimmermann, pp. 208–09. Cf. J. J. Duin, *La doctrine de la providence dans les écrits de Siger de Brabant* (Philosophes médiévaux, III; Louvain, 1954), pp. 223–27.

there does in fact seem to have been an evolution in his position
on the soul and intellect and a notable shifting of ground resulting
in great part from Thomas' attack in the *De unitate intellectus
contra Averroistas*. That evolution can be established especially
from the recently discovered and published questions of Siger on
the *Liber de causis*.[5] The intent of this essay is a brief reexamina-
tion of the interchange that took place between Thomas and Siger
in order to bring out the seriousness with which they treated one
another's views and more especially to underscore the manner in
which Thomas did in fact influence the evolution of Siger's philosophical psychology.

The explicitly psychological works of Siger which are universally accepted as genuine are the *Quaestiones in librum tertium de
anima*, the *Tractatus de anima intellectiva*, and the *De intellectu*.
Siger's *Quaestiones in librum tertium de anima*, which were written sometime between 1265 and 1270, may have been the occasion
for Saint Thomas' *De unitate intellectus contra Averroistas*, which
was issued in 1270. Some of the doctrines contained in them are
also to be found in Stephen Tempier's condemnation of the same
year.[6] The distinguished Italian historian of medieval and Renaissance philosophy, the late Bruno Nardi, proved that Siger's
immediate reply to Thomas' attack was his now lost *De intellectu*,
a work that we know primarily from excerpts and paraphrases to
be found in various works of the Renaissance philosopher, Agostino Nifo. While this reply was probably written almost immediately after the appearance of Thomas' opusculum, that is, in

---

[5] For an affirmation of the evolution, see A. Dondaine and L. J.
Bataillon, "Le manuscrit Vindob. lat. 2330 et Siger de Brabant," *Archivum
Fratrum Praedicatorum*, XXXVI (1966), pp. 206–10; Z. Kuksewicz, *De
Siger de Brabant à Jacques de Plaisance, La théorie de l'intellect chez les
averroïstes latins des XIIIe et XIVe siècles* (Warsaw, 1968), pp. 24–66;
Antonio Marlasca, "La antropología sigeriana en las 'Quaestiones super
librum de causis'," *Estudios filosóficos* (Santander), XX (1971), pp. 3–27.
Gilson, Maurer, MacClintock and Nardi all rejected any evolution, but they
did so before Siger's questions on the *Liber de causis* were brought to the
attention of students of medieval philosophy.

[6] Whether this particular work was under consideration by Thomas
and Tempier would be difficult to establish with certainty. Cf. F. Van
Steenberghen, *La philosophie au XIIIe siècle* (Louvain and Paris, 1966),
p. 435; Edouard-Henri Wéber, *La controverse de 1270 à l'Université de
Paris et son retentissement sur la pensée de S. Thomas d'Aquin* (Paris,
1970), pp. 29–33, 42–45.

1270 or 1271, Siger authored yet another reply, the *Tractatus de anima intellectiva,* which Fernand Van Steenberghen has dated as written in 1272 or 1273.[7] A brief examination of these various works and of Siger's questions on the *Liber de causis,* which were written between 1273 and 1276, will make evident the evolution of Siger's ideas on the soul and the intellect that resulted from his careful study of Saint Thomas' opusculum.

In the very first question of his *Quaestiones in librum tertium de anima,* Siger sets forth the thesis that the intellect is not part of one simple soul along with the vegetative and sensitive parts but is rather a simple being in itself that forms with the vegetative and sensitive parts, once it is united with them, a composite soul (*anima composita*).[8] In the second question, Siger goes on to note that the intellect under discussion, namely, "our intellect" (*intellectus noster*), is not God (*Primum Agens*), but is, according to Aristotle, eternally caused by God, just like the world itself (Q. 2, pp. 5–6). However, the intellect is of its own nature always reducible to nothingness and is perpetually kept in existence solely through the will of God (Q. 5, p. 17). Since God alone is pure act and therefore completely simple and without composition, all other beings recede from his simplicity and have some composition.[9] In the case of the human intellect, which is one for all men (Q. 9, pp. 25–27), this composition does not involve form and matter, not even what some would call "intelligible matter," but rather two parts or elements which are related to one another as matter and form (Q. 6, pp. 18–21). The intellect cannot be directly united by its substance to the bodies of individual men, since it would then be inseparable from the body. It is rather united to the body only accidentally, that is, by its power (Q. 7, pp. 22–24). The intellect

---

[7] Bruno Nardi, *Sigieri di Brabante nel pensiero del Rinascimento italiano* (Rome, 1945), p. 33; Van Steenberghen, *La philosophie,* p. 444. Kuksewicz's proposal (pp. 25, 44, 64) to date the *De intellectu* as later than the *De anima intellectiva* cannot be accepted.

[8] Siger of Brabant, *Quaestiones in tertium de anima,* in Bernardo Bazan (ed.), *Siger de Brabant: Quaestiones in tertium de anima, De anima intellectiva, De aeternitate mundi* (*Philosophes médiévaux,* XIII; Louvain and Paris, 1972), p. 3. Subsequent references will appear in the text.

[9] See also Siger of Brabant, *Questions sur la Métaphysique,* ed. C. A. Graiff (*Philosophes médiévaux,* I; Louvain, 1948), II, Q. 7, p. 21–22; III, Q. 7, pp. 93–96.

exists in the body both by moving the body and also by thinking in the body (Q. 8, p. 25).

Siger now faces the basic problem of how the single intellect, which he also calls the "rational soul" (*anima rationalis*) and the "intellective soul" (*anima intellectiva*), is united in knowledge to individual men and thereby diversified. He tells us in the question devoted to the unity of the intellect that the intellect is numbered or diversified through the different *intentiones* present in the imaginations of different human beings (Q. 9, p. 28; cf. Q. 8, p. 25). The agent intellect and the possible intellect are two powers (*virtutes*) of the rational soul or separate intellect which are involved in our having intellectual knowledge (Q. 15, p. 58; cf. Q. 13, p. 44). The separate intellect has no innate knowledge; it depends rather on phantasms (*phantasmata*), and therefore on human bodies, to gain knowledge. The agent intellect abstracts the universal intelligibles from the *intentiones* or *species* in the imaginations of human beings. These intelligibles are then received by the possible intellect, which thereby knows (Q. 12, p. 40; Q. 14, pp. 46–51; Q. 15, pp. 54–60). Consequently, while remaining one in substance and power, the intellect may be said to be numbered and diversified in different men by reason of their different *intentiones imaginatae*. Siger thus attempts to answer the objection that all men will have the same degree of knowledge if there is only one intellect by maintaining that, since one man will have an *intentio* in his imagination that another does not in fact possess, the former will have the corresponding universal knowledge that the other will lack (Q. 9, pp. 28–29). Nonetheless, basing himself on remarks of Averroës, Siger also claims that the union of the intellect with the human race, which is eternal, is more essential than its union with individual men. The intellect is united essentially with the human race, just as it is related only accidentally to the *intentiones* in the imagination of this individual human being (Q. 14, pp. 52–53; cf. Q. 15, pp. 56 and 58).

When Saint Thomas left Paris for Italy in 1259 or 1260, there was as yet no evident philosophical movement in the Arts Faculty which alarmed him. But during the period from 1260 to 1265 there arose a group in that Faculty which centered around Siger and taught such doctrines as the eternity of the world and the unity of the intellect. It seems likely that one of the major reasons for

Thomas' return to Paris in 1269 was the strength and influence which Siger and his allies had achieved during his absence.[10] Church authorities at Paris eventually proceeded to condemn some of the key philosophical teachings evidently discussed in the Arts Faculty. On December 10, 1270, Stephen Tempier, the bishop of Paris, issued a condemnation of thirteen propositions which concerned four main "errors," namely, the eternity of the world, the denial of universal divine Providence, the unity of the intellect or rational soul for all men, and determinism.[11] However, Siger had already been the object of another far more effective critique, for earlier that same year Saint Thomas had finished and presumably let be circulated his own treatise against the "Averroists," namely, the *Tractatus de unitate intellectus contra Averroistas*, which is a direct and trenchant attack on the psychological ideas of Siger.[12] He singles out as the two fundamental errors of the Averroist interpretation of Aristotle that the possible or potential intellect is not the substantial form of the body and that it is one for all mankind and thus not multiplied according to the number of existing men. Thomas uses passages from Aristotle himself as well as remarks from various Arabs (namely, Avicenna and Algazel) and Greeks (namely, Alexander of Aphrodisias, Theophrastus and Themistius) to show that Aristotle held the possible or potential intellect to be part of the individual human soul, which is the form of the body, and also that it is not just Latin philosophers who attribute this to him (Ch. 1–2, pp. 1–37).[13]

It should be noted that in the preface of the *De unitate intellectus* Thomas charges that those whom he is attacking claim to follow the words of the "Peripatetics" (*Peripatetici*), presumably meaning by that the late ancient Greek commentators, whereas in

---

[10] Angelus Walz, *Saint Thomas d'Aquin*, trans. P. Novarina (*Philosophes médiévaux*, V; Louvain and Paris, 1962), pp. 149–51.

[11] See Pierre Mandonnet, *Siger de Brabant et l'averroïsme latin au XIIIme siècle, IIme Partie*, 2nd ed. (Louvain, 1908), pp. 90–112; Van Steenberghen, *La philosophie*, pp. 375, 472–74.

[12] Saint Thomas Aquinas, *Tractatus de unitate intellectus contra Averroistas*, 2nd ed., ed. Leo W. Keeler (Rome, 1957). References in the text will be to this edition.

[13] Weber (pp. 29–33) has reconstructed from apparent quotes from Siger in the *De unitate intellectus* the basic ideas of another work, one different from the *In tertium librum de anima*.

fact they have never read their works on the subject (Ch. 1, pp. 2–3). Subsequently, in two different parts of the work he quotes at length from the translation of Themistius' paraphrases on the *De anima* which his Dominican colleague, William of Moerbeke, had made for him in order to show that Themistius held that there is an individual possible intellect and agent intellect in each human being (Ch. 2, pp. 33–36; Ch. 5, pp. 77–78). Since Averroës appears to have claimed Themistius as a proponent of the unity of the intellect,[14] Thomas feels emboldened to call Averroës "the perverter of peripatetic philosophy" (Ch. 2, p. 38). He expresses his surprise that some, after having looked only at Averroës' commentary on the *De anima,* would dare state that all Greeks and Arabs hold to the unity of the intellect, especially when Averroës has "perversely" presented the doctrine of Themistius and Theophrastus (Ch. 5, p. 78). William of Moerbeke's translation of Themistius has thus played a crucial role in Thomas' attack on Averroës, and there is good evidence that it also provided him with a new and interesting key to the thought of Aristotle.[15]

As to the unity of the intellect, Thomas attacks only the theory that the possible intellect is one for all men. He sees nothing objectionable in viewing the agent intellect as one for all men—it was in fact identified with God by various thirteenth century thinkers —even though this is not in agreement with Aristotle (Ch. 4, pp. 54–55). Thomas lists at length absurdities which he believes result from the unity of the possible intellect and he argues that such a doctrine is contrary to Aristotle (pp. 55–63). He also takes up and replies to arguments against the multiplicity of the possible intellect, namely, that multiplicity of the intellect could occur only through the division of matter, and this would mean that the intellect would be a material form; that the intelligible object, that is, the essence or quiddity, is the same for all men, and therefore

---

[14] See Averroës, *Commentarium magnum in Aristotelis de anima libros,* ed. F. C. Crawford (Cambridge, Mass., 1953), III, comm. 5, pp. 389–90.

[15] Daniel A. Callus, "Les sources de saint Thomas," in *Aristote et saint Thomas d'Aquin* (Louvain and Paris, 1957), pp. 117–25, 163–70. See also my article, "Themistius and the Agent Intellect in James of Viterbo and Other Thirteenth Century Philosophers (Saint Thomas, Siger of Brabant and Henry Bate)," *Augustiniana* (forthcoming).

the intellect must also be one and the same; that if there remained many intellectual substances after death, they would lack all activity; that there would now be an actual infinity of intellects according to the position of Aristotle, who held to the eternity of the world; and that it is a principle of all philosophers, the Latins alone excepted, that the intellect is not multiplied (Ch. 5, pp. 63–76). In his replies, Thomas develops two crucial doctrines which would much exercise Siger, namely, that of the individuation of the soul and the intellect and that of intelligible species. According to the former doctrine, the intellect is an immaterial power of the soul, and the soul is the act of the body and thus individuated and multiplied by this relationship (*per comparationem*) to matter: there are many souls in many bodies and many intellectual powers in those many souls (pp. 66–67). The latter doctrine admits that the intelligible object (*intellectum*), namely, the universal nature or essence, is the same for all men, but it makes a distinction between the intelligible object which is one and the same for all knowers and the intelligible species (*species intelligibilis*) whereby the individual knows the intelligible object. The agent intellect abstracts the intelligible species from the individual sensible by leaving aside the individual principles or conditions. Consequently, the intelligible species represents a being only according to its universal nature or essence. There will thus be many intellects, each with its own act of knowing and intelligible species, and yet one and the same intelligible object for all of them (pp. 71–72). But perhaps an even stronger weapon that Thomas uses in his attack on Siger and the doctrine of the unity of the intellect is the appeal to personal psychological experience, namely, that it is this individual man who actually knows (*hic homo singularis intelligit*) and not some separate and unique substance (Ch. 3, pp. 39–42). Thomas has thus challenged Siger to explain more adequately the nature of the intellect and its relationship to the human being's seemingly individual and personal act of knowing.[16]

Siger's immediate response to Thomas' *De unitate intellectus* is to be found in his now lost *De intellectu*, which we know pri-

---

[16] For some appreciation of the effectiveness of Thomas' challenge, see Van Steenberghen, *La philosophie,* pp. 432–43.

marily through references in works of Agostino Nifo. The first modern scholar to draw attention to these passages in Nifo, other than Francesco Fiorentino, who made passing allusion to some of them in the last century, was the late Bruno Nardi. The following is part of Nardi's listing of the various doctrines which Nifo relates from Siger's *De intellectu:* 1) the possible intellect is the lowest of the separate substances and is one for all men; 2) while the possible intellect remains one in itself, it acquires a multiplied existence from its relations to different men; 3) the "intellective soul" is produced by the union of the possible intellect with the cogitative power; 4) by reason of this union, the intellective soul can be called the substantial form of man, not just an assisting form, and man receives his specific existence as a rational animal from it; 5) although the agent intellect is God, it too can be called part of the human soul insofar as it concurs in human thinking and also insofar as it is united as the form of the possible intellect at the conclusion of the latter's intellectual development; and 6) man can have a direct knowledge of the separate substances and God during the present life.[17] Oddly enough Nardi does not observe that the meaning of "intellective soul" for Siger has changed here, assuming of course that Nifo is giving Siger's precise terminology. In the *Quaestiones in librum tertium de anima* and in the *Quaestiones de anima intellectiva,* Siger refers to the separate intellect as the intellective soul, whereas according to Nifo's account of Siger's *De intellectu* it means the union of the separate intellect with the cogitative power. Siger evidently held in his *De intellectu* that the vegetative and sensitive parts in the individual man unite as a sort of semi-soul with the separate intellect, which is also taken as a sort of semi-soul. From these two as it were "semi-souls" there results the whole intellective soul. This of course involves a conception much like that of the composite soul set forth in other of Siger's works.[18]

While Siger's shift regarding the agent intellect, namely, to identify it with God, would of course not provoke Saint Thomas' wrath, his reassertion of the unity of the possible intellect for all

---

[17] Bruno Nardi, *Sigieri di Brabante nel pensiero del Rinascimento Italiano* (Rome, 1945), pp. 46–47. Cf. F. Fiorentino, *Pietro Pomponazzi* (Florence, 1868), pp. 246–48.
[18] Nardi, *Sigieri,* pp. 30–31.

men is a deliberate rejection of Thomas' critique in the *De unitate intellectus*. What is interesting, however, is how Siger attempts to come to terms with some of the difficulties which Thomas had set forth. Nifo remarks that Siger was a contemporary of Thomas who wanted to mediate between the Latins and the Averroists (*averroici*) in a treatise which was a reply to that of Thomas. He characterizes Siger as having achieved this mediation by maintaining the undividedness, immateriality, and unity of the intellect with the Averroists, while simultaneously holding with the Latins that the intellect is the form which constitutes man both in his species and also in his individual existence. Siger emphasizes that man is essentially a single composite (*per se unum compositum*) of the body and the potential intellect. Man has his specific difference from the intellect, and through the mediation of the cogitative power, which enables matter to receive the intellect, the being of man as man is also ultimately from that intellect. Siger's argument seems to be that since there is no distinction between the essence of the potential intellect and the potential intellect taken as an individual and singular being, given it is one of the Intelligences, the intellect can give individual existence (*dare esse hoc*) even though it is also a universal essence. Indeed, Siger appears to want to solve the problem of how the single and one potential intellect can give individual existence to many individual human beings by holding that insofar as it is a quiddity or essence with a universal existence it can be divided among different material things suitably informed by cogitative powers, whereas insofar as it is itself an individual being it provides the existence through which a human being is an individual. And yet he also wants to maintain the doctrine which he had already put forward in the *Quaestiones in librum tertium de anima*, namely, that the intellect is primarily and per se the form and act of human nature (*natura*), which is eternal and one for all men, and only accidentally the act or perfection of individual men like Socrates and Plato. It is of course difficult to see how this latter doctrine can be reconciled with the notion of the potential intellect as individual providing individual existence to individual men.[19]

---

[19] Agostino Nifo, *Liber de intellectu* (Venice, 1503), I, tr. 3, ch. 18, f. 32v, 2–33, 1; ch. 26, f. 35v, 2–36, 1. Cf. Nardi, *Sigieri,* pp. 18–20; Van Steenberghen, *La philosophie,* pp. 444–46.

The fact that Siger nowhere in the *De intellectu* surrenders the key Averroist doctrines of his earlier work nor makes any concession to Christian orthodoxy and the strictures of Tempier has led Van Steenberghen to argue quite plausibly that it was written prior to the bishop's condemnation of December 10, 1270.[20] But during the following two or three years, Siger apparently reflected more carefully on Thomas' *De unitate intellectus* and perhaps on arguments contained in other works of Thomas. The result of this philosophical reflection was the work which he completed either in 1272 or 1273, the *Tractatus de anima intellectiva*, which contains modified positions on some of the topics discussed in his earlier works and also reveals very careful attention given to the most difficult of Thomas' arguments in the *De unitate intellectus*.

The most striking feature of Siger's *De anima intellectiva* is that the center of focus has shifted to Aristotle himself. No doubt chafing from Thomas' attempt to discredit his interpretation of Aristotle by citations from "the Philosopher," Siger makes constant appeal to the text of Aristotle in order to rebut Thomas. He also seems eager to refute the charge that he follows Averroës blindly and is unacquainted with the ancient Aristotelians. Indeed, it is revealing that Themistius, the very commentator to whom Thomas had appealed in order to discredit Averroës' reliability as a commentator, is mentioned by Siger most frequently after Aristotle and is even called his "expositor," while Averroës himself is cited on only two occasions.[21] Also striking is the cautiousness that Siger now exhibits, motivated perhaps by Tempier's condemnation of 1270. He is extremely careful to distinguish his task from that of seeking the truth about the soul, since the latter can be achieved only through revelation and not by natural reason. He limits his goal to determining the intention of the philosophers, especially Aristotle, on the soul, and he explains that inasmuch as the philosophers have no direct experience of souls totally separated from the body, they do not teach that souls exist as separate.

---

[20] Van Steenberghen, *La philosophie*, p. 447.
[21] Siger of Brabant, *Tractatus de anima intellectiva*, in Bazan (ed.), *Siger de Brabant* . . . , Ch. 3, p. 83; Prol., p. 70; Ch. 6, pp. 97–98. Subsequent references will appear in the text.

When one proceeds philosophically, one seeks not the truth, but the intentions of the philosophers, looking especially to Aristotle and what is comprehensible to human reason and experience (Ch. 3, pp. 83–84; Ch. 6, pp. 99–100; Ch. 7, p. 101).

Siger attempts to meet Thomas' challenge head on, even while maintaining that according to Aristotle the intellect is one for all men. He explains, first of all, that the soul is the act, form, and perfection of the body, emphasizing like Thomas (*De unitate intellectus*, Ch. 3, pp. 39–40) that that by which something operates primarily and principally is its form (Ch. 2, pp. 72–73). He then attacks both Albert and Thomas by name for holding that while the substance of the soul is united in existence to the body, the soul's power or potency, namely, the intellect, is separate from the body inasmuch as it does not operate through any bodily organ. Siger presents texts to show that such a position contradicts Aristotle, and he argues that according to Aristotle the intellective soul is separate in existence from the body and united to it only operationally, just like a sailor and his ship (Ch. 3, pp. 81–85). But while Siger does still speak of the separate intellect as dependent on the human body and its phantasms in the act of knowing, he now characterizes the intellect as somehow united by its very nature to the body. While the intellect cannot be a true substantial or material form of the body, it can be called its "form" in a broad sense, since whatever is an internal agent in regard to matter (*intrinsecum operans ad materiam*) is its form (Ch. 3, pp. 84–85, 87). By this formula, which involves a rather unclear conception of the intellect's relationship to human bodies,[22] Siger evidently hoped to overcome several of Thomas' crucial objections in the *De unitate intellectus* (Ch. 3, pp. 39–50), namely, that a separate intellect cannot be the form of the body; that the fact that this individual man thinks cannot be explained by the union of his phantasm with the separate intellect; and that if the separate intellect is united to Socrates as a mover to a thing

---

[22] The doctrine is treated almost with contempt by Marcel Chossat, "Saint Thomas et Siger de Brabant," *Revue de philosophie*, XXIV (1914), pp. 573–74. For a more sympathetic account, see Stuart MacClintock, "Heresy and Epithet: An Approach to the Problem of Latin Averroism," *The Review of Metaphysics*, VIII (1954–1955), pp. 187–94, who shows the influence of Siger on Jandun on this point.

moved, Socrates will not be placed in his species through the intellect and his thinking will really belong not to him but to the separate intellect. Siger denies both that man understands by reason of the separate intellect being his "mover" (*motor*) and also that the act of knowing belongs to a man through his phantasms. His position is that the act of knowing can be attributed to the man as well as to the intellect since the operations of internal agents, which are parts of the composites in which they operate, are to be attributed to their composites (Ch. 3, pp. 85–86). In like fashion, Siger insists that this conception of the union of the intellective soul with the body is sufficient to say that man is man through the intellect, that is, that the composite is denominated rational through the presence of the intellect within it. And finally, Siger repeats his view that the intellective soul can be called the perfection and form of the body insofar as it is an agent operating within the body. It is like a true form insofar as that which is intrinsic to a body is not spatially separate from the body, and also because the operation of an internal agent denominates the whole composite.[23] Siger concludes the third chapter of the *De anima intellectiva* with a final jab at Albert and Thomas, pointing out that what Aristotle says of the intellect in Book III of the *De anima*, namely, that it is separated from matter and impassible, holds true not only for the potency of the intellective soul but also for its very substance (Ch. 3, pp. 86–88).

Siger also notes that while the intellective soul is a caused being for Aristotle, it is immortal and eternal from the past as well as into the future (Ch. 4–5, pp. 89–95). However, what Aristotle thought about the nature of the intellective soul's separation from the body Siger judges to be difficult to determine. On the basis of several texts, he argues that Aristotle seems to think that the intellective soul is not wholly separate from all bodies. He points out that according to the exposition of Averroës and *perhaps* (*forte*) also according to Aristotle, while the intellective soul

---

[23] For an excellent discussion of the soul's ability to be "positionally indistinct" from the body, which makes the soul capable of operating within the body, see Charles J. Ermatinger, *The Coalescent Soul in Post-Thomistic Debate* (Ph.D. dissertation: Saint Louis University, 1963), pp. 28–40. Ermatinger shows that Giles of Rome made heavy use of this concept in his own interpretation of Averroës. *Ibid.*, pp. 41–75.

perdures when separated from one particular body which has corrupted, it becomes the act of another body. Siger emphasizes that it is Aristotle's doctrine that both the human species and the intellective soul have always existed; that an infinite number of men have already existed; that the intellect can never lack its activity, for which it stands in need of human bodies; and that an actual infinity of things is impossible. He therefore concludes that it is wholly outside reason and contrary to the intention of Aristotle that individual men could have their own intellective souls which would continue to exist after the corruption of the body, since this would involve that there now exists an infinite number of disembodied intellective souls (Ch. 6, pp. 95–100). But while Siger presents Aristotle here as denying the possibility of a multiplicity of intellective souls, he shows signs of being himself unsure as to what Aristotle, reason, and philosophy hold on the matter and he discusses the question at length.

In the seventh chapter of the *De anima intellectiva*, Siger turns directly to the question whether the intellective soul is multiplied by the multiplication of human bodies, and he states clearly that he will pursue the question philosophically, that is, he will attempt to establish what Aristotle thought and what human reason and experience show. The first three arguments he uses to show that such a multiplication of the intellective soul is impossible are that the intellective soul is separated from matter according to Aristotle and therefore lacks the necessary cause of multiplication of individuals within the same species; that the intellective soul is self-subsistent and individuated in itself and so cannot be common to many individuals; and that plurality results only from quantified being, whereas the nature of the intellective soul is to be separate from quantity (Ch. 7, pp. 101–04). These arguments and further explanatory remarks in this chapter regarding individuation (pp. 105–06) reveal the unmistakable and strong influence of Saint Thomas' doctrine of individuation, an influence first underscored by Franz Bruckmüller over sixty years ago.[24] But Siger also goes on to admit that there are both philosophical authorities and also very difficult arguments in behalf of

---

[24] Franz Bruckmüller, *Untersuchungen über Sigers (von Brabant) Anima intellectiva* (Munich, 1908), pp. 140–53.

the multiplicity of intellective souls. It is striking that the three authorities whom Siger then presents as upholders of the multiplicity of the intellective soul, namely, Algazel, Avicenna, and Themistius (Ch. 7, p. 107), are the very philosophers whom Saint Thomas had used in his *De unitate intellectus* (Ch. 5, pp. 76–77) to discredit the belief that only Latins held to the multiplicity of the intellective soul. Similarly, the two arguments which Siger presents in behalf of the multiplicity of the soul are also inspired by Thomas' opusculum. They are that if the intellect were common to all men, when one man had knowledge, all other men would have the same knowledge, and that if there were only one intellect for all men, the intellect would always be filled with all intelligible species and there would be no need for the agent intellect (Ch. 7, pp. 107–08; cf. Saint Thomas, *De unitate intellectus,* Ch. 4, pp. 57–58, 61). Siger confesses that because of the difficulty of these and other arguments he has long been in a state of doubt (*mihi dubium fuit a longo tempore*) as to what natural reason should hold on the issue and what Aristotle himself believed regarding the question. The only possible course in such a state of doubt is to accept the multiplicity of the intellect on the basis of religious faith (Ch. 7, p. 108).[25]

The uncertainty which Siger expresses in the *De anima intellectiva* regarding Aristotle's position on the unity or multiplicity of the intellective soul reappears in his later questions on the *Liber de causis*. These questions, which were discovered recently and published in a complete and critical edition only two years ago, were probably the result of a course of lectures which Siger gave in the Arts Faculty. Since they show clear dependence on Saint Thomas' own commentary on the *Liber de causis,* which was completed in 1272; disagree with crucial positions which Siger had maintained in the *De anima intellectiva,* written in 1272 or

---

[25] Given Siger's candid admission that he is unsure what is Aristotle's real view on the intellective soul, it is difficult to see how he can be considered a convinced "Averroist" in this work. Furthermore, he has made Themistius (and not Averroës) the major "expositor" of Aristotle in the treatise. Consequently, he would surely have been ill at ease in simply attributing unity of the intellective soul to Aristotle while simultaneously admitting that Themistius maintained the multiplicity in man of both agent and possible intellects.

1273; and yet expound doctrines included in Tempier's condemnation of 1277, they have quite plausibly been dated as having been written sometime between 1274 and 1276, the year that Siger fled Paris.[26]

Siger had evidently continued to mull over the problems and difficulties which Thomas had raised in the *De unitate intellectus* and to which Siger had tried to reply both in his *De intellectu* and also in his *De anima intellectiva*. But he now considers Averroës' doctrine of the unity of the intellect to be not only heretical according to the Christian faith but also to seem irrational. And even while conceding that Aristotle's position on whether the intellect is one or many is not certain from his own words, Siger now declares that whatever Aristotle may have thought on the problem, he was human and could err (Q. 27, pp. 111–12). Siger comes down on the side of the plurality of the intellect and he offers philosophical arguments, not simply an appeal to religious teaching and authority, to justify his acceptance of that plurality. It would thus appear that Siger has abandoned the identification of philosophy and natural reason with Aristotle.

On the fundamental issue of the nature of form, Siger holds that the substantial form of each thing is simple and not composed of more elementary forms or grades of form, showing evident concern for the problem of predicating genus, species, and difference of the same individual. Socrates must therefore be existent, living, and rational through the one simple form (Q. 4, pp. 47–48). Facing the crucial question of whether the human intellect or intellective soul is the form and act of the body, Siger quotes Aristotle (*De anima*, II, 1, 412a27–28) to the effect that soul universally is the act of the body. He then presents three arguments in favor of the thesis which are without doubt inspired by Saint Thomas, namely, that man is determined to his species by the principle of his proper operation, thinking, and since this principle is the intellective soul, it must be his form; that the specific differ-

---

[26] Antonio Marlasca, *Les Quaestiones super librum de causis de Siger de Brabant* (Philosophes médiévaux, XII, Louvain and Paris, 1972), pp. 25–29; Dondaine and Bataillon, pp. 206–07, 210–11. On Siger's flight, see A. Dondaine, "Le Manuel de l'inquisiteur (1230–1330)," *Archivum Fratrum Praedicatorum*, XVII (1947), pp. 186–92.

ence is taken from form, and thinking is man's specific difference and results from the intellective soul; and that since that whereby anything primarily operates is its form, and man's thinking is through his intellect, the latter must be man's form.[27] But while Siger uses these arguments of Thomas to show that the human intellective soul is the substantial form of man, thus abandoning the rather unsatisfying theory of "form" as *intrinsecum operans* which he had put forth in the *De anima intellectiva,* he is still unwilling to adopt Thomas' conception of the soul as united in its essence to matter as the form and perfection of matter while its power, the intellect, is separate from matter and in no way the act or form of the body. It is Siger's own position that both the substance of the soul and its power are the act and perfection of matter, and he argues in behalf of his position that if the power and its operation were in fact completely separated from matter so as to be in no way the act and perfection of matter, then the substance of the soul would also not be the act and perfection of matter. Nonetheless, Siger is quite careful to distinguish the intellective soul as a form of the body from ordinary material forms. He does this by holding that while the intellective soul perfects matter and depends on the body for its object of knowledge, namely, the phantasm, it is a form which is per se subsistent and thus not dependent on matter for existence, that is, as a subject in which it would have its existence. Not surprisingly Siger adds that the intellect is not generated from the potency of matter, as is the case for material forms, but comes from outside (Q. 26, pp. 103–07; cf. Q. 18, pp. 80–81). Siger thus insists, on the one hand, that the intellect communicates with matter (*communicans ad materiam*) to the extent necessary for the multiplicity of human intellects, even calling the intellect "material" at one point, and yet he adds, on the other hand, the important, though perhaps not wholly consistent qualification, that the intellect is not totally material in its existence (Q. 27, p. 113). But perhaps this is not in fact really so distant from Thomas' own thought, since the relation

---

[27] See Thomas Aquinas, *Summa theologiae,* I, q. 76, a. 1; *De unitate intellectus,* Ch. 1, pp. 8–9; Ch. 3, pp. 44–47, 50–51. Cf. Marlasca, "La antropologia sigeriana," pp. 14–15.

with the body that Siger has in mind appears to consist in the intellect's need for phantasms as the object it requires to know anything. And since this dependence of the intellect in its thinking does not take away its subsistence, Siger feels free to argue that the intellect is per se subsistent and an individual thing (*hoc aliquid*) and yet also the act or form of the body (Q. 52, pp. 179–82; cf. Q. 47, p. 169). In like fashion, the theory of individuation which Siger appears to espouse has strong resemblances to that of Saint Thomas. We are told that form is terminated to individual existence through the determinate matter (*determinata materia*) in which it is received, and it is thereby contracted in its being (*in esse contrahitur*) (Q. 53, pp. 183–84; cf. Q. 32, p. 126; Q. 33, p. 127). Indeed, there can be a plurality of human intellects which are specifically the same and yet numerically distinct only because of this union with matter: human souls are multiplied through this relationship (*per comparationem*) to bodies (Q. 27, pp. 112–14, 116). In contrast, any form which is completely separate from matter cannot be multiplied (Q. 18, p. 81; cf. Q. 9bis, p. 59). And while Siger does not discuss at length the problem of the disembodied human soul, he does indicate at one point that when the intellect is separated from the body it will retain the existence which it had in the body, that is, it will continue to have the nature of being the form of a body, and it will therefore also retain its multiplicity (Q. 27, p. 114). This is of course remarkably like Saint Thomas' own solution to the problem of how the disembodied human soul remains individuated.[28]

In the twenty-seventh question, which concerns the central issue of whether the intellect is one for all men or multiplied, Siger presents four arguments against the unity of the intellect which are strikingly reminiscent of ideas of Saint Thomas. They are that bodies which are alike specifically but differ numerically must have numerically distinct souls which agree specifically; that inso-

---

[28] See for example *De unitate intellectus*, Ch. 5, pp. 63–68. On Thomas' doctrine of individuation, see G. M. Manser, *Das Wesen des Thomismus* (Fribourg, 1949), pp. 645–96; Hans Meyer, *The Philosophy of St. Thomas Aquinas*, trans. F. Eckhoff (St. Louis, 1946), pp. 72–80, 176–81; M.-D. Roland-Gosselin, *Le "De ente et essentia" de S. Thomas d'Aquin* (Paris, 1948), pp. 104–34.

far as the intellect is the form of man, and thus one of his intrinsic principles, intellects must be particular and many like human beings; that only if the intellects of men are different can there be differences in the respective knowledges which they possess; and that if there were in fact only one intellect, the act of knowing of Socrates and that of Plato would be identical, which is false.[29] Siger shows philosophical acumen in his diagnosis of why Averroës was led to postulate the unity of the intellect: it was because the Commentator simultaneously believed both that things differ numerically only through quantitatively different matters and also that the intellect is a per se subsistent form which is separate in existence from matter. Moreover, for Averroës the acts of knowing of Socrates and Plato are identical as to their subject and their intelligible object; they differ only by some relative diversity of the intelligible species as it is considered to be caused by the phantasm in Socrates or the phantasm in Plato (Q. 27, pp. 110–12).

Several of the ten arguments in favor of the unity of the intellect which Siger presents in the twenty-seventh question revolve around this key problem of individuation and are not wholly unlike arguments which Siger had himself used in earnest in the *De anima intellectiva*. In his replies to these arguments, Siger makes use of the theory of individuation sketched above and which he undoubtedly culled from Thomas (Q. 27, pp. 108–10, 112–16). But in the ninth and tenth arguments Siger also presents problems relating to the nature of knowledge, the intelligible species and the intelligible object. The ninth argument is that if the intelligible species were received into an intellect individuated by matter, the intelligible species would be a principle only of knowledge of the singular and not of individual knowledge. And the tenth argument is that if Plato and Socrates did have numerically different intellects, then the intelligible objects that each possesses when it knows the same essence or nature will be numerically different and they will both still stand in need of some common and abstract intelligible object; the intelligible object will thus be in fact numerically singular, not universal, and it will be only potentially

---

[29] See Thomas Aquinas, *Summa theologiae*, I, q. 76, a. 2; *De unitate intellectus*, Ch. 4, pp. 55–58; *Q. D. de spiritualibus creaturis*, a. 9. Cf. Marlasca, "La antropologia," pp. 20–21.

intelligible and not actually intelligible (Q. 27, p. 110). These arguments are doubtless inspired by discussions of Saint Thomas in his *De unitate intellectus* and other works, but so too are the solutions which Siger works out.[30] Siger thus observes in reply to the ninth argument that the intelligible species involves abstraction from the individual conditions of a thing and that the intellect is not individuated through matter in such fashion that the intelligible species are received in some physical organ or in matter. And in his reply to the tenth argument he adopts wholesale the distinction which Thomas had made central in his critique in the *De unitate intellectus,* namely, that Averroës failed to distinguish the intelligible object (*intellectum*), which is one and the same for many knowers, from the intellects (*intellectus*) and their intelligible species (*species intelligibiles*), both of which are many. That is to say, Plato and Socrates have different intellects and different intelligible species whereby they know one and the same intelligible object or universal nature (pp. 110, 116–17).

The theory of knowledge which Siger puts forth in the questions on the *Liber de causis* also shows the mark of Saint Thomas' influence, though there are some nuances which seem to be Siger's own. For one thing, Siger nowhere speaks of the potential or possible intellect and the agent intellect as potencies or powers that are really distinct from the soul itself, as Thomas does in some of his writings. And while Siger does not explicitly state that the possible and agent intellects are simply the soul functioning in different ways, he does at one point argue that the human intellect is not bound in its existence to matter, since the nature of this intellect extends itself universally to all material forms by both abstracting and receiving (Q. 52, p. 181).[31] The human intellect is in potency to receive all universal forms by means of

---

[30] See Thomas Aquinas, *Summa theologiae*, I, q. 76, a. 2, ad 4; *De unitate intellectus,* Ch. 5, pp. 68–73; *De spiritualibus creaturis*, a. 9, ad 6. Cf. Marlasca, "La antropologia," p. 25, n. 66.

[31] Marlasca ("La antropologia," p. 14) appears to identify the human intellect solely with the possible intellect and he suggests that the agent intellect is a somewhat unclear entity whose nature cannot be established from the texts of the *Quaestiones*. I believe that there is nothing mysterious about the agent intellect; it functions in the human soul as that which abstracts the intelligible species, just as the possible intellect functions by receiving those species.

intelligible species (Q. 22, pp. 93–94; Q. 30–31, pp. 122–23; Q. 39, pp. 148–49; Q. 45, pp. 161–62). However, our intellect is by its nature united to the body and dependent on phantasms for it to know anything whatever; this is, as was already noted, a dependency as on an object and not as on a subject in which it exists (Q. 48, p. 149; Q. 52, p. 182). Indeed, both the agent intellect and the potential or possible intellect require phantasms and the activities of the sense powers, the one to have something from which to abstract the intelligible species, and the other to have something to receive (Q. 37, p. 146). The phantasm and the agent intellect thus together reduce the possible intellect from potency to act (Q. 41, pp. 151–52; cf. Q. 52, p. 178). Material forms are the connatural and proportionate intelligibles for our intellect, but they are known by our intellect only because it has abstracted them from their individuating principles (Q. 48, p. 170). While the sensible species is a principle whereby we know both a thing and all its individuating conditions, the intelligible species which results from the abstraction of these conditions leads to universal knowledge (Q. 28, p. 116).

From this review of Siger's pre-1270 *Quaestiones in librum tertium de anima,* and then of his *De intellectu, De anima intellectiva,* and questions on the *Liber de causis,* all of which were written after Saint Thomas' *De unitate intellectus,* the evolution of Siger's psychology has been established. The various arguments and authorities which Thomas had presented forced him to reconsider over a period of several years both his interpretation of Aristotle and also what he himself thought human reason must conclude regarding the nature of the intellective soul. We have seen how Siger first replied to Thomas' challenge with a defensive readjustment of Averroist themes in the *De intellectu.* While he also continues to offer arguments for the unity of the intellect in his slightly later *De anima intellectiva,* he now expresses doubt as to whether Aristotle had indeed held to the unity or multiplicity of the intellect and what natural reason should maintain on the issue. Moreover, evidently impressed by the effective use which Thomas had made of Themistius to discredit the reliability of Averroës as a commentator, Siger substitutes Themistius for Averroës as his major guide to Aristotle. Finally, in the questions on the *Liber de causis,* he states that Aristotle's position on the unity or multiplicity of the intellect is not certain, but whatever he thought, he was still human and fallible. In the work, philosophical arguments are set forth for the plurality of the intellect which show the clear influence of Saint Thomas, and the views on individuation and the nature of cognition which Siger presents show the same influence. However, Siger still refuses to accept Thomas' distinction between the essence of the soul and its power, namely, the intellect. The influence of some of Saint Thomas' ideas on the evolution of Siger's psychology should not be taken to mean that Siger simply became a Thomist.

The philosophical interchange between Thomas and Siger that we have studied here centered on the soul and the intellect, but there is also evidence of Thomas' influence on Siger in other areas as well.[32] Indeed, extant writings of other members of the Arts Faculty dating from the 1270's show heavy study of and dependence on the writings of Saint Thomas.[33] The respect of the masters of that faculty for him is also evident from their corporate letter of condolence to Thomas' Dominican confreres on the occasion of his death in 1274.[34] Siger and his colleagues learned from Thomas and they valued both his respect for the Philosopher and also his concern for the rights of philosophy vis-à-vis reactionary forces at Paris—basic doctrines of Thomas would be condemned in 1277 by Stephen Tempier, bishop of Paris, along with doctrines of masters of the Arts Faculty like Siger and Boethius of Dacia. In turn, there is some evidence that Thomas' debate with Siger regarding the intellect did have important re-

---

[32] The influence on Siger of Saint Thomas' conception of theology in the *Summa theologiae*, I, q. 1, has been demonstrated recently by William Dunphy and Armand Maurer in their brief but important article, "A Promising New Discovery for Sigerian Studies," *Mediaeval Studies*, XXIX (1967), pp. 364–69. For another example, see R.-A. Gauthier, *Magnanimité* (Paris, 1951), pp. 475–77.

[33] O. Lottin, "Saint Thomas d'Aquin à la faculté des arts de Paris aux approaches de 1277," *Recherches de Théologie ancienne et médiévale*, XVI (1949), pp. 292–313.

[34] Alexander Birkenmajer, "Der Brief der Pariser Artistenfakultät über den Tod des hl. Thomas von Aquino," in *Vermischte Untersuchungen zur Geschichte der mittelalterlichen Philosophie* (*Beiträge zur Geschichte der Philosophie des Mittelalters*, XX, 5; Münster, 1922), pp. 1–35. Cf. Wéber, p. 19.

percussions in his own psychology and theory of knowledge, and he may even have decided to write commentaries on all the works of Aristotle as a response to Siger's challenge.[35] Thomas was

---

[35] Edouard-Henri Wéber presents as the basic theme of his book, *La controverse de 1270 a l'Université de Paris et son retentissement sur la pensée de S. Thomas d'Aquin* (Paris, 1970), that a serious philosophical dialogue did in fact take place between Thomas and Siger, one that had its effects not only on Siger but also on Thomas himself (pp. 15-25). By beginning the technique of literal commentary on Aristotle, which was to characterize the Arts Faculty, Siger forced the problem of the intellect into the open around 1266 and provoked other masters in the university to discuss the thorny topic. The most important of these responses was of course Saint Thomas' *De unitate intellectus contra Averroistas* and we are thus in Siger's debt for having led Thomas to penetrate more deeply into the problem of the intellect in Aristotle and to have further developed his psychology and theory of knowledge in some remarkable ways (pp. 36-37). Wéber argues that it was Siger's challenge and Thomas' eagerness to justify Aristotle in the face of the conservative theologians' irritation with Siger's teaching that prompted Thomas' decision to write commentaries on all the works of Aristotle. He was aided in this project by Moerbeke's translations of Aristotle and the Greek commentators, especially Themistius and Philoponus (pp. 42, 112-13, 116). Wéber discerns two basic doctrinal developments in Thomas which resulted from the interchange with Siger, the one regarding the relation of the soul and intellect and the other regarding the intelligible object. The first development, which is perhaps the more debatable, is that Thomas adopted Siger's view that the intellect is the very substance of the soul. Wéber argues that as a result of the controversy Saint Thomas abandoned the notion of a real distinction between the essence of the soul and its operative powers, which involved that the powers were considered as "accidental forms" of the soul (pp. 87-91, 98-109). In place of this essence-power-operation distinction, Thomas adopted the notion of a principle and its operation. That is to say, the principle by which we know is the intellect and it is identical with the intellective soul, which is the form of the body (pp. 117-27). While the intellect is a power (*potentia*) of the soul, it is also the first act of man, the form of the human body (pp. 142-54, 195-208). Wéber argues that as a result of Saint Thomas' effort to combat Siger's dualistic conception of the intellect as the "mover" (*motor*) of the body, Thomas adopted some ideas of the pseudo-Dionysius and came to conceive of the potential intellect as the animating principle of man, that is, as his substantial form (pp. 209-20). The second development is that of Thomas' notion of intelligible species, abstracted from phantasms, which are distinct from the intelligible object. He uses this doctrine to undercut Averroës' argument that since all men have the same intelligible objects they must have the same intellect. Thomas broke the Aristotelian identity of intellect and intelligible which Averroës follows by introducing the intermediary of a *similitudo* in human cognition, namely, the intelligible species. He thus defeated Averroës by elaborating, with the aid of the pseudo-Dionyius, a doctrine which runs counter to Aristotle's own conception of the basic identity in cognition of the intellect and the intelligible object (pp. 221-56, 278-79, 289-91).

very much a man of his day and we shall never fully appreciate his philosophical achievement if we do not develop a thorough understanding of the philosophical challenge presented to him by his contemporaries. Siger was surely the most impressive and acute philosophical mind with which Thomas had to contend, and the respect he had for the Brabantine master is evident from the thorough and careful replies that he addressed to him in the *De unitate intellectus* and other works. It is fitting that Dante immortalized in his *Paradiso* the intellectual bond between these two great Parisian masters.

Duke University.

Oddly enough, Wéber appears to have some doubts regarding the ascription of the questions on the *Liber de causis* to Siger, but he offers no reasons for these doubts (pp. 129, 218). I hope to examine Wéber's book more thoroughly on another occasion. While many of his ideas and theses are appealing, and the documentation is impressive, his tendency to distinguish early and late strains in the same work (e.g. the *Summa theologiae* and the *Summa contra Gentiles*) in the style of Werner Jaeger will perhaps demand further examination and evaluation. In any case, his book is an important contribution that must be studied by all interested in either Thomas or Siger.

THE PROBLEM OF THE EXISTENCE OF GOD IN SAINT THOMAS' *COMMENTARY ON THE METAPHYSICS* OF ARISTOTLE

FERNAND VAN STEENBERGHEN
Translated by JOHN WIPPEL

IN A RECENT ARTICLE I have attempted to indicate the contribution of St. Thomas' *Commentary* on Aristotle's *Physics* to the study of the philosophical problem of God's existence.[1] Here I propose to undertake a similar investigation with respect to his *Commentary on the Metaphysics*. The method will be the same as that followed in the previous article. The first part presents a sketch of the "theology" of Aristotle as it is developed in his books on "first philosophy." The second part determines how Thomas Aquinas himself viewed his role as a commentator on Aristotle. Is his exegesis faithful and objective? Does he respect the Philosopher's historical positions? Does he stress the differences between the Aristotelian Pure Act and the Christian creative God? Or does he rather attempt to reduce the distance between these two to a minimum? It is enough to raise these questions for one to appreciate their historical and doctrinal interest.

I

It is well known that Aristotle's *Metaphysics* presents historians with many problems of literary history. The fourteen books of "first philosophy" available today did not originally form one single work which Aristotle himself planned and authored. This work as we have it is rather the result of a compilation after its author's death. His disciples have joined together different writings which reflect their master's teaching at different points in his career. Consequently, historians are confronted with problems of chronology, both absolute and relative,

---

[1] F. Van Steenberghen, "Le problème de l'existence de Dieu dans le commentaire de saint Thomas sur la *Physique* d'Aristote," *Sapientia* (La Plata) 26 (1971), pp. 163–72.

with respect to each of the parts they distinguish in our version of the *Metaphysics*. One's understanding of Aristotle's thought and of the development of his teaching depends in large part on the solution to these problems.

No attempt will be made here to enter into these problems, nor even to summarize the current positions of Aristotle's major historians. Moreover, these questions of literary criticism have little importance for our topic, since Thomas Aquinas himself neither raised nor even suspected them. He always regarded the *Metaphysics* as a treatise whose composition was homogeneous, and even managed to find a logical structure in it.

Aristotle's *Metaphysics* also raises a host of problems of interpretation for contemporary exegetes. These are of more direct interest to the main theme of this article. In order to move to a comparative study of the views of Aristotle and of Thomas Aquinas, one must begin by establishing the authentic positions of the Stagirite himself. Once again, however, even a brief resume of recent controversies as to the precise nature of the "theology" of Aristotle would go far beyond the limits of this article. I will restrict myself to recalling the main points that are in dispute, and then will set forth the Philosopher's "theology" by drawing upon the surest and most widely admitted conclusions of contemporary scholarship.

The first problem to be discussed has to do with the object and the unity of that science which Aristotle calls "first philosophy." In what sense is it first? Is it first because it studies the first substances, the separate substances, and in particular, the most perfect of these, the First Mover? If so, it would deserve the title "divine science" or "theological science." But will it still be the universal science, the science of being as being, the science of substance in general? One can show, it would seem, that the supreme theoretical science, the study of being as being and of substance in general, should also treat of separate substances and especially of the first of these, and thus that it ends in a "theology."

But here a new problem arises. What is the meaning of the terms "God" and "divine" in Aristotle's philosophy and what is their origin? It appears to be certain that in Aristotle's work philosophical investigation of the question of God takes its point of departure from Greek mythology and the popular religion that it produced. In the Stagirite's eyes, these mythological and popular conceptions veil a mysterious and profound reality which one must recover and express in philosophical language.

Finally, modern interpreters differ considerably as to the precise meaning of a number of passages in the *Metaphysics* which bear upon the Aristotelian "theology," especially when one compares them with occasional allusions and remarks found in other works of the Stagirite. Some of these difficulties will be singled out in what follows.[2]

An attempt will now be made to summarize the more surely established points of Aristotle's "theology" as it is found in the *Metaphysics*. Since Book XII contains the only systematic exposition of this "theology," this analysis will begin with the views developed there.

In the *Physics* the Stagirite had demonstrated that one must acknowledge the presence of a First Mover, perfectly unmoved, unique and eternal, at the origin of all movement found in the universe. But he viewed this First Mover as immanent within the corporeal world, even though it itself is immaterial. The First Mover is the soul of the first thing that is movable, that is to say, of the first heaven, the sphere which encircles all the other heavenly spheres. Moreover, he saw in this First Mover the *efficient cause* of the circular movement that it communicates to the first movable thing by exercising its causal influence on the latter's periphery.[3]

---

[2] The contemporary literature on problems raised by Aristotle's *Metaphysics* is vast. I will limit myself here to citing some recent works which can serve as a point of departure for a more thorough investigation. For works prior to 1962 the essentials are to be found in the excellent *Note bibliographique* joined to the French translation of D. J. Allan's *The Philosophy of Aristotle* (Oxford, 1950): *Aristote le Philosophe* (Louvain, 1962). See in particular Sections I (Vie, évolution, doctrine d'Aristote) and III (Métaphysique), pp. 229–32. From 1962 onward see: P. Aubenque, *Le problème de l'être chez Aristote* (Paris, 1962); "Sens et structure de la métaphysique aristotélicienne," *Bulletin de la Société française de philosophie* 58 (1964), n. 1 (Séance du 23 mars 1963); G. Reale, *Aristotele. Le Metafisica. Traduzione, Introduzione e Commento* (Naples, 1968); Ch. Lefèvre, *Sur l'evolution d'Aristote en psychologie* (Louvain, 1972). Some references are also to be found in J. C. Doig, *Aquinas on Metaphysics. A Historico-doctrinal Study of the Commentary on the Metaphysics* (The Hague, 1972), Part Two, pp. 99–340.

[3] Cf. F. Van Steenberghen, "Le problème de l'existence de Dieu . . . ," pp. 165–66.

The views developed in Book XII of the *Metaphysics* are rather different. Here the First Mover is no longer described as the soul of the first sphere or as the efficient cause of its movement. It is absolutely separate from anything sensible whatsoever. Pure act, free from any connection with matter and potency, it is not an efficient cause of movement. For this would imply a certain dependence in it with respect to movement. It is rather its final cause. It moves the world in the manner of an object that is loved which draws toward itself beings whose natural appetite is directed toward its perfection. It is an immaterial substance and is at one and the same time the First Intelligible and the First Intelligence. It is Thought which thinks itself, Thought always in act. That is why it is a perfect, eternal, and immutable substance. It knows nothing but itself, not because of any defect or imperfection, but because there are things which it is better not to know. A knowledge of these would debase the knowing subject. Finally, it possesses in surpassing fashion that delight which goes with thinking.[4]

In Book XII, chapter 8, Aristotle discusses other immaterial substances and tries to determine their number in the light of data drawn from astronomy. There must be as many immaterial moving substances as there are irreducible heavenly movements. But, if it follows from his treatment that these substances are subordinated to the First Mover just as the things they move are subordinated to the first movable thing, Aristotle never gives an exact indication as to the nature of this subordination and the nature of the relations which obtain between the First Substance and the others.

Chapter 10, the final chapter of Book XII, presents interpreters with some difficult problems. Here the Philosopher considers the order of the universe and the relationships of beings therein with one another. He compares the universe to an army where the soldiers' activities are directed to the good of the leader; then to a family where all members of the community, children,

---

[4] On all of this see chapters 6, 7, and 9 of Book XII. For a good resume of the evolution of Aristotle's "theology" from his dialogue *On Philosophy* to Book XII of the *Metaphysics* see S. Decloux, *Temps, Dieu, liberté dans les commentaires aristotéliciens de saint Thomas d'Aquin* (Bruges, 1967), pp. 157–65.

servants, even domestic animals, are ordered to the common good because of the presence of the father of the household, who is the principle of unity and coordination. After criticizing the views of Empedocles and Anaxagoras, he concludes to the necessary unity of a Principle of universal order, which moves all beings as their supreme Good. At first sight, this discussion contradicts chapter 9, where any idea of "providence" seems to be excluded because Pure Act can only know its own perfection, and where knowledge of the universe is regarded as debasing. Some commentators conclude from this that chapter 10 is an earlier treatment in which Aristotle presents a view closer to that of Plato. Others eliminate the difficulty by minimizing the importance of the comparison of the First Mover with the leader of an army or the father of a household.

Apart from Book XII there are other passages in the *Metaphysics* which can be seen as allusions to "theological" problems and which will be understood in that way by Saint Thomas. Let us attempt to indicate more precisely their importance in Aristotle's work.

While discussing truth at the beginning of Book II, Aristotle declares that being to be supremely true which is the cause of truth for "posterior" beings. This is why, he continues, it is necessary for the principles of beings which always exist to be supremely true. Indeed, there is never a time when they are not true. They have no cause for their being, rather they cause the being of other things.[5] What are these eternal beings to which Aristotle here refers? How is one to understand the "principles" of such beings? These principles are uncaused, but they serve as causes of being for others. In what way are they causes of being? In Aristotle's system the heavenly spheres and their immaterial movers always exist. It is therefore likely that in this passage from Book II the Stagirite is designating these immaterial movers as "principles," that is to say, the first beings, the first substances, in the order of eternal beings. These immaterial substances are causes of being for the heavenly spheres, not in the strict sense of creative causes, but insofar as they produce their movement and

---

[5] *Met.* II, c. 1 (993b 24–31).

thus give to them the "moved being" in which their ultimate and proper perfection consists.[6]

At the beginning of Book VI Aristotle places the science of being as being at the peak of the theoretical sciences, for it studies beings that are separate (from matter) and immovable. Then he adds: "But it is necessary that the common (that is, universal) causes be eternal, and especially these (that is, those which are separate from matter), for they are causes of things that are manifest from the world of sensible beings."[7] This, then, is another allusion to the moving substances of the heavenly spheres, which are eternal, for they exercise permanent causality on the lower world (by means of the spheres).

In Book X, while considering unity, Aristotle shows that unity exists in all genera and that a principle of unity should be sought in each one. After giving different examples from the realms of quality and quantity, he adds that the same holds for the genus of substance. One must here seek a substance which is the principle of unity (in the order of substances).[8] One might think that the Stagirite here refers to the substance of the First Mover. Insofar as it is at the peak of the order of substances, it is a principle of unity within that order, with all others taking their place in the hierarchy of substances by reason of their relation to that First Substance. But this interpretation is not admitted by all commentators.

Such, then, is the data provided by the *Metaphysics* on its author's "theology." In spite of his lofty concept of the First Cause, Subsisting Thought and Perfect Beatitude, Center of Attraction for the Whole Universe, Aristotle never arrived at the idea of creation. The First Cause is a cause of movement, that is to say, of change in all its forms, beginning with the local circular

---

[6] This passage cannot be understood as implying creative causality, for such an interpretation would contradict Aristotle's entire philosophical system. On this see A. Mansion, "Le Dieu d'Aristote et le Dieu des chrétiens," *La philosophie et ses problèmes. Recueil d'études de doctrine et d'histoire offert à Monseigneur R. Jolivet* (Lyons-Paris, 1969), pp. 21–44. See pp. 23–25.

[7] *Met.* VI, c. 1 (1026a 15–18). The translation used by Saint Thomas is not an exact rendering of the Greek text. It should read: "but it is necessary for *all* (first) causes to be eternal, but especially these, for they are causes of that which, *among things divine*, is manifest (to the senses)."

[8] *Met.* X, c. 2 (1054a 11–13).

movement of the heavenly spheres, the most perfect movement of all. The First Cause is not a cause of being or a creative cause. This is not to say that the Philosopher expressly rejects the notion of creation. But he did not see the problem of metaphysical causality and thus does not raise the question as to the origin of finite beings. His "God" is a finite God (for Aristotle the infinite is the undetermined, the imperfect, the incomplete), who does not know the lower world and does not exercise any kind of personal providence with respect to it. Finally, this First Mover is only the most perfect of a series of immaterial movers of the heavenly spheres. One might be tempted to say that it is simply *primus inter pares*. Its relationships with other immaterial substances are no more precisely indicated than is its relationship with the human intellect. In short, it is difficult to find a true theism in Aristotle's system, for his First Mover is neither creator, nor infinite perfection, nor truly transcendent, nor provident, nor unique absolute, for it shares aseity with other immaterial substances, with the heavenly bodies, and even with the matter of the sublunar world.

II

Saint Thomas' *Commentary* covers the first twelve books of the *Metaphysics*. The question of this work's date is complicated and has been studied by different scholars. There is a complete treatment of this research in a recent work by J. C. Doig.[9] It is probable that there were two successive redactions of certain books of the *Commentary*. In that case the first could go back to Thomas' period in Rome (1265–1267). In any event the definitive redaction of this *Commentary* dates from his final period at Paris. Begun in 1270, it was finished in the beginning of 1272. It is therefore later than the *Commentary on the Physics* which seems to have been put in circulation around 1268.[10]

The *Commentary on the Metaphysics* has not yet appeared in the Leonine edition of Thomas' works. I will therefore cite the convenient Turin edition (Marietti), published in 1915 by M-R.

---

[9] *Aquinas on Metaphysics* . . . , pp. 10–22 (on the chronology of the *Commentary's* composition).

[10] Cf. F. Van Steenberghen, "Le problème de l'existence de Dieu . . . ," p. 163.

Cathala and reedited in 1950 by R. Spiazzi.[11] This is not a critical edition but is "eclectic" as Cathala puts it (p. VIII), in that the text was established by drawing on the earliest editions, in particular the first Venice edition (1503), the Piana (1570), two other Venice editions (1588 and 1593), and the Parma edition (1866). Aristotle's text is given in a medieval Latin translation, fairly close to that which Thomas used beginning with Book V.[12]

The study which follows has profited from many recent works, in particular an article by A. Mansion, a work by S. Decloux, and especially J. Doig's monograph.[13] In this investigation I will follow the order of the *Commentary*. The reason leading me to begin my study of Aristotle with Book XII does not apply here.

Let us first turn, therefore, to the passage from Book II where the Stagirite states that "the principles of beings which always exist are uncaused, but they are causes of being for other things." Saint Thomas comments on this passage in Lectio 2 (295–296). He interprets the Aristotelian formulae in the strongest sense. The principles of the heavenly bodies (he clearly has in mind the separate substances which move the heavenly spheres) are true to an eminent degree. This is so, first of all because they are *always true*, in contrast with beings subject to generation and corruption; again, because they are *uncaused*, in contrast with heavenly bodies. The latter, while being incorruptible, nonetheless have a cause not only of their motion as some have maintained, but also of their being, as the Philosopher himself expressly says here.[14]

---

[11] M-R. Cathala, *Sancti Thomae Aquinatis ... In Metaphysicam Aristotelis commentaria* (Turin, 1915)/R. Spiazzi, *In duodecim libros Metaphysicorum Aristotelis expositio* (Turin, 1950).

[12] Doig notes that the Latin version of Aristotle printed in the Cathala-Spiazzi edition is not the *Metaphysica Moerbecana*, but a combination of the *Media* and the *Moerbecana*, closer in fact to the *Media* than to the *Moerbecana* (*Aquinas on Metaphysics ...*, p. 7 and note 5). On Thomas's use of different translations of the *Metaphysics* in his *Commentary*, see pp. 4–10 of the same work.

[13] A. Mansion, "Le Dieu d'Aristote et le Dieu des chrétiens," (see note 6); S. Decloux, *Temps, Dieu, liberté* (see note 4); J. Doig, *Aquinas on Metaphysics ...* (see note 2).

[14] The essential passage reads: "*Secundo quia nihil est eis (=principiis corporum caelestium) causa, sed ipsa sunt causa essendi aliis. Et per hoc transcendunt in veritate et entitate corpora caelestia: quae, etsi sint incorruptibilia, tamen habent causam, non solum quantum ad suum moveri, ut quidam opinati sunt, sed etiam quantum ad suum esse, ut hic Philosophus expresse dicit*" (Cathala, n. 295).

Thomas then develops his thought by presenting a proof for the radical dependence of heavenly bodies. All composed and "participating" beings must be traced back to beings (plural!) which are by reason of their very essence as to their causes. But all corporeal beings are beings in act insofar as they "participate" in certain forms. Consequently, a separate substance, form by its very essence, must be the "principle" of such corporeal substance.[15]

This passage of his *Commentary* is a typical instance where Thomas takes advantage of a certain ambiguity in Aristotle's text when it is considered in isolation (without taking into account his whole system). Thomas' esteem for Aristotle is so great that he finds it difficult to believe that the greatest of the philosophers was not aware of one of the most fundamental metaphysical truths, creative causality. All the more so since, in the cultural climate of the thirteenth century, the existence of God as unique creator of the universe was in the eyes of all a truth easily discovered and universally recognized. Given this, atheism must have appeared to the medievals as an unusual position, unworthy of a true metaphysician, and to be accounted for only because of a strange intellectual blindness. It was quite unreasonable to think that the Philosopher had fallen into such a gross error.

Thomas is therefore on the watch for the smallest indication that might disclose the presence of the idea of creation in Aristotle. Such is the situation with the passage from Book II. Since Aristotle declares that the separate substances are "causes of being" he wishes to say that they create the heavenly bodies. Thomas explains this fundamental dependency by means of the doctrine of participation.

Not only does Thomas attribute the doctrine of creation to Aristotle on insufficient grounds, but his *Commentary* is disturbing for other reasons as well. The doctrine which he exposes is at mid-point between the authentic view of the Philosopher and the

---

[15] "*Necesse est ut omnia composita et participantia reducantur in ea quae sunt per essentiam sicut in causas. Omnia autem corporalia sunt entia in actu inquantum participant aliquas formas. Unde necesse est substantiam separatam, quae est forma per suam essentiam, corporalis substantiae principium esse*" (Cathala, n. 296). On this demonstration see Doig, *Aquinas on Metaphysics* . . . , pp. 284–88.

authentic notion of creation. First he speaks, following Aristotle, as though there were *many uncaused beings:* "*The principles* of the heavenly bodies are uncaused and are causes of other beings." Then the composition and participation which manifest the dependence of the creature are explained in terms of *formal participation.* Separate substances are causes of substances composed of matter and form because they (the former) are *forms by their very essence.* As one can see, there is no question here either of constitutive composition of the finite being (*esse-essentia*), or of *Esse subsistens,* the *unique* cause of all composite beings.

This same troubling attitude reappears in Book VI, Lectio 1 (1164), where Thomas comments on the passage singled out above, referring to "common causes." The first causes of beings subject to generation should themselves be ungenerated, especially those that are absolutely immobile and immaterial (here he obviously has in mind the moving substances of the spheres). And he adds: "For these immaterial and immobile causes are the causes of the sensible beings that we perceive, because they are beings to the maximum degree and, therefore, causes of others, as Aristotle has shown in Book II. From this it follows that the science that studies these beings is the first of all and considers the *common causes of all beings.*"[16]

This passage marks progress over the one from Book II. Now participation is located in the order of being and no longer in the order of forms. Immobile and immaterial beings are causes because they are *maxime entia*. This implies that their effects participate in being to a limited degree. But the strange plural is still there. There is still question of eternal cause*s,* as if many *maxime entia* and many *causae essendi* could exist. To be sure, it is Aristotle's text which accounts for this plural, but it is surprising that his commentator does not express any reservations on this point after having presented the Philosopher's thought. This does not prevent Thomas from concluding by rejecting as

---

[16] "*Hae namque causae immateriales et immobiles sunt causae sensibilibus manifestis nobis, quia sunt maxime entia, et per consequens causae aliorum, ut in secundo libro ostensum est. Et per hoc patet quod scientia quae huiusmodi entia pertractat, prima est inter omnes et considerat communes causas omnium entium*" (Cathala, n. 1164).

false the view of those according to whom Aristotle's God (singular, this time!) would not be the cause of the substance of the heaven but only of its motion.[17]

In Lectio 3 of Book X, Thomas comments on the Aristotelian doctrine of unity. Here he hardly goes beyond the thought or the letter of the Stagirite. With Aristotle he recognizes that one must seek for a certain substance in the order of substances to which unity belongs essentially. This will be first and foremost the case of that which is first in the order of substances, as Aristotle will subsequently determine.[18] It is clear that for Thomas the "first substance" is God himself who possesses unity as an essential attribute.

One could raise the objection that Thomas places God in the genus of substance. He would undoubtedly reply that God is not *in* the genus of substance, properly speaking, but that he is the "First Substance" because he possesses preeminently the prime characteristic of every substance, that is, to subsist in itself (*ens in se*). He is not in any way in the other categories, the accidents, for their common characteristic is to "inhere" in substance (*ens in alio*). To inhere in another is imperfection by its essence, since it implies dependence. In short, substance is a simple perfection and can therefore be analogically attributed to the creative Cause.

We are now in position to move to Thomas' commentary on chapters 6–10 of Book XII where Aristotle's "Theology" is developed. Thomas' commentary on chapter 6 is remarkably faithful, but he does add some personal observations including some criticism of Aristotle. He first notes that in the *Metaphysics* Aristotle is convinced of the eternity of motion and of time. Otherwise he would not have based his demonstration of the existence of separate substances on the same. He observes that the arguments offered in Book VIII of the *Physics* in support of the eternity of motion are only probable; unless they are simple *ad hominem* arguments intended to refute the views of the ancient physicists

---

[17] "*Ex hoc autem apparet manifeste falsitas opinionis illorum qui posuerunt Aristotelem sensisse quod Deus non sit causa substantiae caeli, sed solum motus eius*" (Cathala, n. 1164).

[18] Cathala, nn. 1972–73. On this text see Doig, *Aquinas on Metaphysics*, pp. 288–90 and in particular the interesting note on p. 290.

(the PreSocratics). Likewise, the proof for the eternity of time developed here, that is, in Book XII of the *Metaphysics,* is not demonstrative. Thus if one supposes that time began to be (together with the real movement of the universe), any duration prior to that beginning is nothing but *imaginary* time, like the space that we imagine beyond the first heaven. Finally, in spite of the inadequate nature of these proofs for eternity of motion and of time, the eternity and the immateriality of the First Substance stand as necessary conclusions. Indeed, if the world is not eternal, it had to be produced by some preexisting cause. If that in turn is not eternal, it too must depend on a cause. And since one cannot regress to infinity, one must stop with an eternal substance that is without potentiality of any kind and, therefore, immaterial.

It will be noted that in Thomas' commentary as in the Aristotelian text there is question at times of *one* eternal and immaterial substance, and of *many* at other times. Aquinas here stays very close to the text on which he is commenting and does not appear to be concerned with disassociating himself immediately from the views expressed therein. That is not to say, of course, that he accepts them all.

In Book XII, Lectio 7, Thomas' commentary clarifies in striking fashion the difficult text of the Stagirite, and does so without forcing his thought. Here our exegete explains why only the final cause moves without being moved. An efficient cause of movement is always subject to movement in some way. If it is a *natural* moving cause, it cannot move without itself being involved in the change. If it is a *voluntary* moving cause, it is moved by the object that is willed. Thomas then shows that, on the human level, that which moves insofar as it is desirable is not always that which moves as an intelligible good. Man can desire an apparent good, the object of a lower appetite. In God, however, the first object understood and the first object loved are one and the same. Farther on he observes that, if the First Mover is the end of the universe, it is clearly not the kind of end that is to be brought into being but rather the kind that preexists, in which lower beings desire to participate (2528). At the end of this Lectio he makes an observation that does not seem to be entirely in conformity with Aristotle's thought. Aristotle, says Thomas, holds that the necessity of the first movement is not absolute but relative to its end,

God. But God is a willing and intelligent being, as Aristotle will go on to show. It follows that any necessity of the first movement depends on the divine will.[19]

Lectiones 9 and 10 comment on chapter 8 of Aristotle's text, where there is question of other separate substances and their number. Thomas faithfully follows Aristotle's text, but does add some personal developments to his commentary. He recalls that astronomers after the time of Aristotle discovered the movement of the fixed stars and therefore had to add a sphere that surrounds that of the fixed stars. He then cites the role of the First Intelligence in Avicenna's metaphysics and criticizes his view that there can only be one immediate effect of God. Thomas casts doubt on the Aristotelian view according to which the hierarchy of separate substances would be parallel to that of movements and things moved. He then takes up the role of the different planets. But all of that is rather removed from our theme.

Thomas' commentary on chapter 9 (Lectio 11) is important, for here he concerns himself with determining precisely in what sense God is the only object of his own thought. He himself knows himself perfectly and by himself. But in knowing himself he also has perfect knowledge of his effects, the heaven and the whole of nature. Thomas adds that the low status of an object of knowledge does not detract from the dignity of the knowing subject, so long as knowledge of such lower things does not prevent knowledge of higher things.[20]

Finally we turn to Lectio 12, commenting on chapter 10. In that chapter Aristotle considers the First Mover as the principle

---

[19] *"Attendendum est autem quod, cum Aristoteles hic dicat quod necessitas primi motus non est necessitas absoluta, sed necessitas quae est ex fine, finis autem principium est quod postea nominat Deum, inquantum attenditur per motum assimilatio ad ipsum, assimilatio autem ad id quod est volens et intelligens, cujusmodi ostendit esse Deum, attenditur secundum voluntatem et intelligentiam, sicut artificiata assimilantur artifici inquantum in eis voluntas artificis adimpletur, sequitur quod tota necessitas primi motus subjaceat voluntati Dei"* (Cathala, n. 2535).

[20] *"Nec vilitas alicujus rei intellectae derogat dignitati. Non enim intelligere actu aliquod indignissimum est fugiendum, nisi inquantum intellectus in eo sistit, et dum illud actu intelligit, retrahitur a dignioribus intelligendis. Si enim intelligendo aliquod dignissimum etiam vilia intelligantur, vilitas intellectorum intelligentiae nobilitatem non tollit"* (Cathala, n. 2616).

of universal order. Once more Thomas casts light on Aristotle's concise text in admirable fashion. But he sometimes goes beyond Aristotle's thought in emphasizing the providential role the latter seems to attribute to the First Mover when he compares it to the head of an army or the father of a household. Without apparent concern for the contradictions which his interpretation of chapter 10 establishes between it and the preceding one where the immanence of the First Mover is presented as a magnificent kind of solitude that excludes any kind of providence, Thomas explains that the order of the universe is the unfolding of a plan which exists in the understanding and will of the First Mover.[21] Farther on he concludes his reflections on nature by observing that, if natural beings act for an end without being aware of it, this is because they receive from the First Intelligence an inclination that directs them towards that end.[22] Finally, he interprets the last phrase of the chapter so as to find therein an express affirmation of the universal providence of God, the unique Principle of the order of the universe.[23] In short, one finds in this last *lectio* views which recall the Fifth Way of the *Summa theologiae*.

III

In bringing this study to a close, some conclusions can now be drawn.

1) The "God" which Aristotle reaches in the *Metaphysics* is certainly superior to the First Mover of the *Physics*. It is an immaterial substance, the most perfect of all, and completely separate from matter. As First Intelligence and First Intelligible, Thinking on its own Thinking, and Pure Act, it moves the universe in the manner of an object of love. But it does not create the uni-

---

[21] "*Totus enim ordo universi est propter primum moventem, ut scilicet explicatur in universo ordinato id quod est in intellectu et voluntate primi moventis. Et sic oportet quod a primo movente sit tota ordinatio universi*" (Cathala, n. 2631).

[22] "*Ex hoc patet quod res naturales agunt propter finem, licet finem non cognoscant, quia a primo intelligente assequuntur inclinationem in finem*" (Cathala, n. 2634).

[23] "*Et hoc est quod concludit, quod est unus princeps totius universi, scilicet primum movens, et primum intelligibile, et primum bonum, quod supra dixit Deum, qui est benedictus in saecula saeculorum. Amen*" (Cathala, n. 2663).

verse. It does not know lower beings, nor does it exercise any kind of providence over them.

2) Saint Thomas does not note any contradiction in Aristotle's texts: neither between the *Physics* and the *Metaphysics,* nor between the last two chapters of Book XII of the *Metaphysics.*

3) The literal interpretation which he gives to the texts in question is almost always accurate and, as in all his commentaries, illuminating. He manages to render perfectly comprehensible the most obscure pages of "this desperately difficult work," as W. D. Ross has expressed it.[24] But we have found two passages where he takes advantage of ambiguity in the text in order to find therein the doctrine of creation, when in fact this doctrine is completely foreign to Aristotle's philosophical system. Again, he interprets a remark by the Stagirite in the sense of an affirmation of divine freedom. Finally he interprets chapter 10 of Book XII in such fashion as to find therein a notion of providence that does not coincide with Aristotle's views.

4) On many occasions Thomas joins to his commentary either personal observations or criticisms of Aristotle or of Avicenna. He does not hesitate to differ with Aristotle on matters such as the eternity of the world and divine knowledge. Since these digressions are clearly distinguished from his commentary properly speaking, they do not compromise its objectivity. In other cases Thomas simply explains the text of Aristotle without singling out the errors that it contains, especially when Aristotle affirms plurality of eternal and uncaused substances or speaks in the plural of *maxime entia.*

5) As to the philosophical problem of demonstrating God's existence, the *Commentary on the Metaphysics* contributes nothing new to the doctrine Thomas has developed in earlier writings. One finds here again the essential elements of the First Way (XII, Lectiones 5–11), some views that recall the Fourth Way (II, Lectio 2 and VI, Lectio 1) and the Fifth Way (XII, Lectio 12). But these discussions contain nothing that is really new.

The Catholic University of Louvain.

---

[24] W. D. Ross, *Aristotle's Metaphysics* (Oxford, 1924), I, p. vi.

AQUINAS ON THE TEMPORAL RELATION BETWEEN CAUSE AND EFFECT

WILLIAM A. WALLACE

It would be difficult to find views more markedly opposed regarding the temporal relation between cause and effect than those propounded respectively by Hume and Aristotle. For Hume, cause and effect must be temporally distinct; in fact, almost by definition the cause is an event anterior in time to the effect, another event. The key problem of causation in Hume's formulation is thus one of making precise the way in which these two temporally distinct events are connected, and he decides this by opting for psychological projections into reality rather than conceding the existence of necessary connections in nature. Aristotle, on the other hand, accepts the fact of productivity or causal efficacy in nature, and so defines cause as to have it actually causing only when it is producing an effect, and thus at best instantiated when cause and effect are simultaneous. While not rejecting the possibility of antecedent causation, in the sense of denying outright that the cause might somehow temporally precede the effect, Aristotle treats this possibility as troublesome and as contributing little or nothing toward the understanding of basic causal processes.[1]

Contemporary thinkers who address the problem of causal relations generally favor Hume's analysis, although some periodically manifest interest in Aristotle's exposition as an important and viable alternative. Few, however, find among the many philosophers who came between Aristotle and Hume any worthwhile contributor to the development of this problematic. Some might note, for example, Nicholas of Autrecourt as a medieval precursor of Hume, but this merely keeps the discussion fluctuating between the same two poles. This essay aims to call attention to a differ-

---

[1] For the basic texts and a discussion of the problematic and its history, see J. S. Wilkie, "The Problem of the Temporal Relation of Cause and Effect," *British Journal for the Philosophy of Science* (1950), pp. 211-29.

ent and intermediate view, not hitherto noted, that was proposed in the High Middle Ages by Thomas Aquinas. It argues that Aquinas made a significant advance beyond Aristotle in his analysis of antecedent causation, and thereby made possible the certification of some elements in Hume's analysis, without subscribing to its more extreme results. In so doing, moreover, Aquinas adumbrated some problems in contemporary analytic discussions of the causal relationships between events, and consequently may shed light on their solution.

I

Aquinas' contributions to natural philosophy and to the methodology of science are not so well known as is his work in metaphysics, but they proved considerable nonetheless. A number of his distinctive teachings in these fields, in fact, had important consequences for the origin of modern science.[2] For purposes of subsequent exposition, we shall here restrict ourselves to one such teaching that runs through his physical writings and that gives rise to a peculiar methodological problem. This is the theme that the *scientiae naturales* are concerned uniquely with changeable being, or with being in process, which process pertains to the sensible or phenomenal order, and on this account is readily available for empirical observation and analysis. The processes that are thus studied by the natural philosopher, however, unfortunately have a contingent aspect to them, in the sense that they are not absolutely necessitated but could be otherwise than they are. If this is the case, then the possibility of attaining true demonstrative knowledge of natural processes would seem to be compromised, and the *scientiae naturales* might have to forfeit their claim to being sciences *simpliciter* in the sense of the *Posterior Analytics*. Aquinas was quite aware of this seeming incompatibility of contingent process and necessary demonstration, and nonetheless maintained that both features serve to characterize the *scientiae naturales*. In maintaining this, moreover, he was led to develop

---

[2] A brief sketch of these contributions is given in my article on Aquinas in the *Dictionary of Scientific Biography*, Vol. 1. (New York: Charles Scribner's Sons, 1970), pp. 196–200.

a distinctive theory of proof for the natural sciences, which may be his greatest single innovation as a scientific methodologist.[3]

As regards the scientist's primary concern with the analysis of process, perhaps the following brief indications may serve to establish this general Thomistic thesis, itself quite Aristotelian in character. The first two books of the *Physics,* in Aquinas' view, are seminal for all of natural philosophy, for they serve to delineate respectively the principles of changeable being and the principles of natural science.[4] Consequent on the determination of these principles, the remaining six books of the *Physics* are devoted to a study of the general properties of process or change.[5] And the subsequent books in the Aristotelian corpus, *De caelo, De generatione et corruptione, Meteorologica,* etc., investigate in detail the various types of process found in the physical universe, viz., local motion, alteration, and growth. From these latter treatises, as is well known, were developed the modern sciences of physics, chemistry, biology, etc., which first separated themselves from the main body of philosophy and then underwent independent development. But even the last six books of the *Physics,* still generally regarded as philosophical in character, concentrate on process as their most formal concern. Thus the third book, which for Aquinas contains the first strict demonstration in natural science, establishes that change is properly found in the thing undergoing change and not in whatever initiates it, and that infinitude, insofar as it is studied by the natural scientist, is formally connected with process insofar as the latter is lodged in some way in the sensible continuum.[6] Demonstrations in Book IV show that every process takes place in time, in Book V that process is possible

---

[3] This theme is developed at some length throughout my two-volume study, *Causality and Scientific Explanation* (Ann Arbor: University of Michigan Press, 1972-74); see Vol. 1, pp. 71-80, 102, 104, 143, and Vol. 2, pp. 247, 250, 293, and 354.
[4] *In II Physicorum,* lect. 1, n. 1.
[5] *In III Physicorum,* lect. 1, n. 1.
[6] There has been no satisfactory full-scale study of the various demonstrations that are to be found in Aristotle's physical works; for a preliminary outline, indication of sources, etc., see my "Some Demonstrations in the Science of Nature," *The Thomist Reader 1957* (Washington, D.C.: The Thomist Press, 1957), pp. 90-118.

strictly speaking only in categories of being that allow contrariety (location, quality, and quantity), and in Book VI that everything that undergoes a process of these kinds must be a divisible or quantified body. The last two books, finally, propose to demonstrate the existence of a First Unmoved Mover, and do so, not through a metaphysical analysis of potency and act, but rather through an analysis of the requirements for initiating change in a body that has the capability of successive movement in time.[7]

The concern with spatial and temporal succession that runs through the *Physics* and the subsequent *scientiae naturales* suggests that Aquinas will also address himself to the problem of the successive and temporal relationships between cause and effect, and this does in fact prove to be the case. The principal loci for this treatment are his commentary on the second book of the *Physics,* toward the end, where he is discussing causal analysis and its mode of employment in the natural sciences, and in the commentary on the *Posterior Analytics,* where these problems are addressed more pointedly as jeopardizing the possibility of demonstrative proof when treating of any natural process.[8] In brief, for Aquinas no special problem is presented when analyses are made in terms of intrinsic (i.e., formal and material) causality, but serious difficulties present themselves when one attempts to demonstrate through extrinsic (i.e., efficient and final) causality, particularly when treating of the sublunary world, for here efficient causes can be impeded from attaining their normal effects. Aquinas' distinctive solution is to propose that demonstrations in the *scientiae naturales* can circumvent the defective operation of efficient causes, whether these arise through material defects or "on the part of time alone," through a technique known as demonstrating *ex suppositione finis*.[9] This technique begins by studying natural processes and noting how they terminate for the

---

[7] For some details, see my "The Cosmological Argument: A Reappraisal," *Proceedings of the American Catholic Philosophical Association* (1972), pp. 43–57.

[8] *In II Physicorum,* lect. 11–15; *In II Posteriorum Analyticorum,* lect. 9–12.

[9] *In I Posteriorum Analyticorum,* lect. 16, n. 6; *In II Posteriorum Analyticorum,* lect. 7, n. 2 and lect. 9, n. 11; *In II Physicorum,* lect. 15, n. 2; see also the references given in note 3 above.

most part. Thus, in biological generation, it is easily noted that men are normally born with two hands, or that olive plants are usually produced from olive seeds provided that these are properly nurtured. From this information, however, one cannot be certain in advance that any particular child will be born with two hands, or that each individual olive seed will produce an olive plant. The reason for this is that the processes whereby perfect organisms are produced are radically contingent, or, stated otherwise, that generating causes do not always attain their effects. But if one starts with an effect that is normally attained, he can use his experience with nature to reason, on the supposition of the effect's attainment, to the various antecedent causes that are required for its production. It is this possibility, and the technique devised to assure it, that permit the *scientiae naturales* to be viewed as sciences in the strict sense. They can investigate the causes behind natural phenomena, they can know with certitude how and why effects have been produced, and they can reason quite apodictically to the requirements for the production of future effects, even despite the fact that nature and its processes sometimes fail in their *de facto* attainment.

To illustrate this technique the favored example of Aquinas, taken over from Aristotle and previous commentators, is the causal analysis of a lunar eclipse. Such eclipses do not always occur, but when they do occur they are caused by the earth's being "diametrically interposed between sun and moon."[10] Thus, if a lunar eclipse is to take place, this will require a certain spatial configuration between sun, moon, and the observer on earth. A similar contingent occurrence is the production of the rainbow in the atmospheric region of the heavens; this is more difficult to explain than the lunar eclipse, since it lacks even the regular movements of the celestial spheres to guarantee its periodic appearance. In fact, rainbows are only rarely formed in the heavens, and sometimes they are only partially formed; when they are formed, moreover, they seem to come about as the result of a contingent process. This notwithstanding, they can still be the subject of investigation

---

[10] *In I Posteriorum Analyticorum*, lect. 16, n. 6.

within a science *propter quid,* if one knows how to go about formulating a demonstration in the proper way. Rainbows do not always occur, but they do occur regularly under certain conditions; they are not always fully formed, but for the most part they form a circular arc across the heavens. An observer noting the regularity of this phenomenon can rightly expect that such a regularity has a cause, and he may proceed to discover what that cause may be. If he moves scientifically, according to Aquinas, he will take as his starting point the more perfect form that nature attains regularly and for the most part, and using this as the end or final cause, will try to discover the antecedent causes that are necessarily entailed in its realization. The necessity of his reasoning is therefore *ex suppositione finis,* namely, based on the supposition that a particular end is to be attained by a natural process. *If* rainbows are to occur, they will be formed by rays of light being reflected and refracted in distinctive ways through spherical raindrops successively occupying predetermined positions in the earth's atmosphere with respect to a particular observer.[11] The reasoning, though phrased hypothetically, is nonetheless certain and apodictic; there is no question of probability in an argument of this type. Such reasoning, of course, does not entail the conclusions that rainbows will always be formed, or that they will necessarily appear as complete arcs across the heavens, or even that a single rainbow need ever again be seen in the future. But if rainbows *are* formed, they will be formed by light rays passing through spherical droplets to the eye of an observer in a predetermined way, and there will be no escaping the necessity of the causal operation by which they are so produced. This process, then, yields scientific or epistemic knowledge of the rainbow, and indeed it is paradigmatic for the way in which the *scientiae naturales* attain truth and certitude concerning the contingent matters that are the proper subject of their investigations.

---

[11] The details of this mechanism were not known to Aquinas but were discovered shortly after his death by another Dominican who had studied at Paris, Dietrich or Theodoric of Freiberg, whose contribution to optical science is discussed in my article in the *Dictionary of Scientific Biography,* Vol. 4 (1971), pp. 92–95. For a full analysis of Theodoric's optical methods, see my *The Scientific Methodology of Theodoric of Freiberg* (Fribourg: The University Press, 1959).

II

The foregoing may serve to show that Aquinas' discourse, with its concentration on natural processes that take place successively and in time, allows for the possibility of temporal intervals between cause and effect. Unfortunately his examples of the eclipse and the rainbow, concerned as they are with light rays, obscure the point somewhat, since in the Middle Ages light propagation was generally regarded as instantaneous. From other examples, however, one can be assured of Aquinas' awareness of the possibility of a time lag between cause and effect. Thus it comes as no surprise to find him treating this point explicitly in his commentary on the *Posterior Analytics,* and there registering an advance over Aristotle's analysis. The locus is chapter 12 of Book II, where Aristotle treats problems relating to the inference of past and future events, and where he raises the question whether there are, as experience seems to show, causes that are distinct in time from their effects. The answer Aristotle proposes is somewhat ambiguous; consistent with his teaching in the *Physics,* he allows that one may infer the occurrence of an earlier event from that of a later event, but denies that the inference can be made the other way around. Most of his discussion then bears on the latter impossibility, where he argues that, were a later event to be inferred from an earlier, during the interval between them it would not be true to say that the later event either has happened or will happen. Here he likens the two events to the points terminating a line segment and the interval or process between them to the line itself. Two such events cannot be either continuous or contiguous, any more than two points can be, nor can the intervening process be contiguous with either event, any more than a line can be contiguous to a point. Thus there is nothing that can serve to hold events together and so assure that any coming-to-be will actually follow upon a past event. Moreover, there always seems to be the possibility of some third event intervening between the two being considered, which could be the cause of the later event's production and therefore would render the inference invalid.

This line of argument has proved troublesome for many commentators, who question Aristotle's identification of cause and effect with point-like events and wonder why he never considered the possibility of cause and effect being more similar to processes, which then could be considered as successive or continuous.[12] It is in evaluating this possibility that Aquinas' commentary proves helpful, for Aquinas does consider the latter case and indeed makes use of it in coming to a solution. In fact he devotes three *lectiones* to an exposition of this one chapter in Aristotle, and in the first two of these even attempts to show how, in accordance with the Stagirite's principles, a cause that is not simultaneous with its effect may still serve as a middle term in a demonstration.[13] The argument Aquinas uses to support his interpretation is relatively simple: just as in any process prior and subsequent elements can be identified, so in the causal processes by which natural agents produce their effects prior and subsequent elements may similarly be noted. Or, as he puts it,

> since the notions of prior and subsequent are necessarily involved in any process, in considering the causes of a process one must accept the fact that the cause and the thing caused are likewise related as prior and subsequent. For it is obvious that a natural agent cause produces its effect through some type of process; and just as the thing undergoing change is brought to the terminus of the process through the entire process itself, so through the first portion of the process it is brought to the second portion, and so on. Hence, just as the entire process is the cause of the subsequent state of rest, so the first portion of the process is the cause of the subsequent portion, and so on.[14]

Aquinas then goes on to note that this line of reasoning is applicable whether one object alone is undergoing change or whether a series of such objects are acting upon one another successively:

> This analysis is true whether it is confined to one object that is in process without interruption from beginning to end, or is applied to several objects the first of which initiates change in the second, and the second in the third. And although, while the first in the series effects change in its object at the same time as the object itself under-

---

[12] See, for example, W. D. Ross, *Aristotle's Prior and Posterior Analytics*, a Revised Text with Introduction and Commentary (Oxford: Clarendon Press, 1949), pp. 80–81 and 648–53; and Hugh Tredennick, *Aristotle's Posterior Analytics*, The Loeb Classical Library (Cambridge, Mass.: Harvard University Press, 1960), pp. 13–15 and 219–27.
[13] *In II Posteriorum Analyticorum*, lect. 10–12.
[14] *Ibid.*, lect. 10, n. 2.

goes change, nevertheless the object thus changed continues to initiate change in another object even after it ceases to be changed itself. In this way several movable objects undergo change successively, with the one being the cause of the change induced in the other, and so on. . . .[15]

The example Aquinas supplies here is that of the projectile, which, though faulty from the viewpoint of modern science, was readily understood by his contemporaries; the point would be better illustrated in our day with the propagation of water waves by a stone dropped into a mill pond, for the case of the single object, and with the successive falls of a row of dominoes, for the case of the plurality of objects. In such instances, as in the example Aquinas supplies,

. . . the cause is not simultaneous with that of which it is the cause [i.e., the effect], insofar as the first portion of the process is the cause of the second, or the object first undergoing change induces a change in the second.[16]

Having therefore conceded the possibility of antecedent causality, or of a temporal interval between cause and effect, both of which he likens to the parts of a process, Aquinas then takes up the more difficult question as to how antecedent causes can serve as middle terms in scientific demonstrations. To clarify his exposition he introduces an example that is particularly apposite in that it expands the time interval between cause and effect considerably beyond that noted in cases involving the transmission of light rays. The example is that of a person taking medicine and subsequently being cured, either at some specified time, such as "he will be cured on such and such a day," or at some unspecified time, such as simply "he will be cured in the future."[17] In terms of this illustration it is easy for Aquinas to explain why Aristotle has reservations about syllogizing from an earlier to a later event, and why he seems to countenance the possibility of the reverse procedure. Once a person has been cured, it does seem reasonable to attribute the cause of his cure to his prior taking of the medicine. But from the fact that a person takes medicine at a particular time, one may not infer scientifically that he will

---

[15] *Ibid.*
[16] *Ibid.*
[17] *Ibid.*, nn. 8 and 9.

be cured either at some specified later date or indefinitely in the future. The reason for this is that, as Aristotle indicates, after taking the medicine there will always be some intervening time "in which it is true to say that he had drunk the medicine but not yet true to say that he has been cured," [18] or, and this is Aquinas' emendation, that "having posited what is prior, the subsequent does not necessarily follow in cases where the effects of the causes can be impeded." [19]

A difficulty yet remains, however, and this relates to the type of syllogizing to which Aristotle apparently has given approval, namely, that of drawing an inference from a later to an earlier event. Here too, in the intervening time, it would seem always possible to find an intermediate event, different from the taking of the medicine, that could yet serve as the cause of the person's cure. If this possibility exists, then one can never be sure that he was cured by the taking of the medicine, and thus scientific knowledge would seem to be precluded even through this *a posteriori* reasoning process.[20]

The problem, of course, already exists for Aristotle, but the solution he devises is not at all clear, and one suspects that this is why he prefers to insist on the simultaneity of cause and effect whenever necessary demonstrations are required.[21] Here Aquinas' commentary again proves helpful, for the Common Doctor attempts to meet the objection and thus still guarantee the possibility of demonstrating in natural science through antecedent causality. Aquinas admits that if one conceives of events as point-like completions of processes, there will always be an infinite number of such completions or partial completions, and on this basis alone it will be impossible to know where to start or to terminate in any demonstrative process. The practical problem is not insoluble for him, however, for he notes that one can always begin with the point-like event that corresponds to the moment "now," and from this reason back to a cause that is ultimately

---

[18] *Ibid.*, n. 8.
[19] *Ibid.*, n. 9.
[20] *Ibid.*, lect. 11, n. 5.
[21] Cf. *ibid.*, lect. 10, n. 9.

immediate with respect to the process that produces the noted effect. To illustrate this he extends the example of the person taking the medicine to include a further process consecutive on his being cured, namely, "his performing the tasks of a healthy man." [22] Should the sick person now be observed performing such tasks, one can reason back that it was necessary for him to have been cured at some time previously, and if he has been so cured, it is further necessary that he previously have taken the medicine. Thus, if D stands for "performing the tasks of a healthy man," C for "being cured," and A for "taking medicine," C can function as a *causa cognoscendi* that serves to connect D with A as its ultimate cause. On this basis, writes Aquinas,

> we can conclude that if D has come to pass, it is necessary that A have previously come to pass; and we take as cause that which took place in the interim, namely, C. For D having come to be, it is necessary that C have previously come to be; and C having come to be, it is necessary that A have previously come to be; therefore, D having come to be, it is necessary that A have previously come to be. For example, if this person has now performed the tasks of a healthy man, it follows that previously he had been cured; and if he was cured, it is necessary that previously he had taken the medicine.[23]

The foregoing is merely illustrative of the procedure Aquinas would recommend to pass through various mediate events or processes until one finally is able to demonstrate the immediate cause of the effect being investigated. So he continues:

> Therefore, by always taking a middle in this way, for example, something else between C and A, as C was taken as middle between D and A, one will come to rest somewhere at something immediate.[24]

This is as far as Aquinas goes with that particular example, but perhaps its further consideration in the light of present-day knowledge may serve to clarify his point. Let us assume, in Aquinas' example, that the person who takes the medicine is incapacitated by severe stomach acidity, and that the medicine he drinks is some form of alkalizer. If one is to believe television commercials, the essential mechanism of the resulting cure will be provided by some intermediate process or event, which may be

---

[22] *Ibid.*, lect. 11, n. 4.
[23] *Ibid.*
[24] *Ibid.*

designated as *B*, and which will consist in a chemical reaction whereby the alkali ions neutralize the acid ions in the sick person's stomach, and thereupon gradually restore him to health. Thus understood, the entire process of the cure may be seen as made up of four partial processes: *A*, the ingestion of the medicine; *B*, the neutralization of the stomach acidity; *C*, the restoration of normal functioning to the other organs of the body; and *D*, the performance of the tasks of a healthy man. All of these components, it may be noted, are themselves processes, although they begin and terminate with point-like events, up to and including the moment "now" from which the reasoning process started. From the viewpoint of modern medicine, what is most important is that the proper or immediate cause of the cure is best seen microscopically as the event-like combination or neutralization of the alkali and acid ions, where for each particular combination the particles come into contact, and where, in this micro-process, partial cause and partial effect are themselves simultaneous. But such individual micro-processes aside, the entire neutralization process is *not* simultaneous, being made up of a series of such micro-processes taking place over a time interval, and the same is true of the taking of the medicine and the final effecting of the cure, both of which depend on the movement of medicine and the redistribution of organic fluids, and thus are time-consuming processes.

On an understanding such as this, based on Aquinas' recommended method, it would seem possible to have scientific knowledge of processes wherein efficient causality is exercised over a period of time, and where the initiating cause is temporally antecedent to the completed effect. The methodology of demonstrating *ex suppositione finis* then can be seen as applying to such cases, just as it does to the cases of the lunar eclipse and the production of the rainbow. Some medicines, it is true, prove to be ineffective, and some people are not cured—a state of affairs completely analogous to that in the example, based purely on nature's operation, that not every human being is born with two hands. But such defects arising either from matter or from the fact that nature acts over an extended period of time, while complicating the methodology whereby demonstrations can be attained, in no way nullify the possibility of scientific or epistemic knowledge of nature and its processes.

III

This much said, some differences between the Humean and the Aristotelian views of causality may now be clarified in the light of Aquinas' commentary. Hume, it would appear, was accurate in his intuition that the exercise of efficient causality, as observable at the phenomenal level and so of particular interest to the scientist, would involve sequential series of events wherein the cause would generally be apprehended prior to the effect. Once he had decided on an event-like analysis of cause and effect, moreover, he was correct in maintaining that such an analysis can never yield knowledge of necessary connections in nature. Aristotle's difficulty in analyzing cases where cause and effect are punctiform, with the infinite number of possibilities they provide between the extremes of any natural process no matter how short, already signals the conclusion to which Hume would be forced once he had restricted himself to a consideration of atomic events alone. On the other hand, Hume's limitation lay in being too precipitate when urging that the only meaningful analysis of causation would have to remain at the phenomenal level, and there invoke merely an event ontology. The experience of recent science has shown, for example, the poverty of such dogmatic empiricism for providing knowledge of the entities and mechanisms that underlie observable events and that now serve to reveal their actual connectedness.

Aristotle, as has been observed, is enigmatic with respect to the problem considered in this essay. One could maintain, and indeed some Aristotelian Thomists would be so inclined, that the interpretations here attributed to Aquinas are actually those of Aristotle himself. That this is unlikely may be seen from an examination of the commentaries on the Greek text of the *Posterior Analytics,* and also from a study of the major commentators in the Latin West, including Averroës, Robert Grosseteste, Albert the Great, and extending even to Jacopo Zabarella. Aquinas' interpretations may be seminal in Aristotle, but their distinctive articulation is not to be found elsewhere in the commentatorial tradition. With regard to the *ipsa verba* of Aristotle, it should be noted that he himself suggested event-like analyses as appropriate for dealing with problems of antecedent causality, and thus is partly responsible for the difficulties into which an exclusive concern with such analyses would later lead. The reason for this suggestion is probably associated with his view of substantial generation or change, which takes place at an instant of time and therefore is best described as a point-like event. Since such substantial change is the normal terminus of many of the natural processes studied in the *scientiae naturales,* it is not surprising that punctiform events should have emerged large in Aristotle's thought. In his favor, however, it should be noted that he himself was quick to realize the difficulties inherent in event analysis, and possibly for this reason consistently de-emphasized antecedent causality, preferring rather to discuss cases where cause and effect are simultaneous. Such a preference, as it turns out, was methodologically sound, for the reduction of causal processes to their immediate initiators at the micro-level brings one ultimately to instantiations of simultaneous causality. The search for such deeper levels of explanation also takes one beyond the phenomenal order to regions of ontological depth where otherwise hidden mechanisms can be explored and the ways in which these serve to connect phenomenal events made apparent.

Aquinas, as we have attempted to show, combines both Humean and Aristotelian elements in his treatment of causal processes. The cases of temporal succession that interest Hume and that led to his causation doctrine were clearly of interest to Aquinas also. The ambiguity in Aristotle as to whether some causes actually do precede their effects or merely appear to do so, is resolved by Aquinas in a way that legitimizes the temporal-succession aspect of Hume's analysis. Moreover, in view of Aquinas' interest in process, and then considering the subsequent development of modern science with its pervasive spatio-temporal descriptions of physical events, it is probable that Aquinas would have admitted that antecedent causality *is* more frequently encountered in the investigation of natural processes that interest the scientist. On the other hand, Aquinas's empiricism could not restrict itself to a Humean form of phenomenalism, but would use the succession of observable events as a springboard to search for their deeper underlying connections. In such a search his sympathies would be with Aristotle and the ultimate resolution of antecedent causation to cases of simultaneous causality. Otherwise the advantage of Aquinas' "middle of the road" view is that it provides a framework in which both ways of studying causal processes acquire a certain legitimacy. The phenomenal method, with its stress on the regularity of succession of events in temporal sequences, and thus in its concern with antecedent causation, serves to isolate instances of natural phenomena whose study may lead to scientific or epistemic knowledge, whereas the more noumenal method, with its search for causal efficacy, underlying mechanisms, and micro-processes that lay bare the connections between events, and this in terms of simultaneous causality, itself results in the sought-after knowledge.

Finally, contemporary analytic discussions of causal relations are not without relevance to the problematic under discussion. For the most part analytical philosophers have given up on Hume's insistence that temporal sequence is essential to causation and its recognition, and are now willing to countenance the view that cause and effect, even when seen uniquely as events, can be simultaneous.[25] Recently, however, an attempt has been made to argue "for a qualified endorsement of Hume's intuition," by showing that there must always be a time difference between cause and effect for the cases "in which cause and effect are modifications of the same physical object."[26] The case proposed is rather peculiar and, although discussed with considerable analytical acumen, seems to have been contrived mainly to accommodate a growing body of what the author regards as "non-problematic" statements concerned essentially with event ontology.[27] It would take us too far afield to survey this literature and evaluate the various moves made therein for the description and recognition of events,[28] but the general impression one gets is that it shows little awareness of the actual problems one encounters when employing causal reasoning in scientific explanation. Much more attractive than this Humean exercise are the proposals of other recent writers

---

[25] For example, Arthur Pap, *Elements of Analytic Philosophy* (New York: The Macmillan Company, 1949), pp. 220–24.
[26] Carl G. Hedman, "On When There Must Be a Time-Difference Between Cause and Effect," *Philosophy of Science* (1972), p. 507.
[27] *Ibid.*, pp. 507, 510.
[28] Hedman gives a brief bibliography on p. 511.

who question the adequacy of any event ontology to deal with causal processes, and turn instead to an investigation of generative mechanisms and structures at the micro-level that explain natural phenomena in terms of the factors that can actually account for their production.[29]

It would be too much to say that all of this problematic has been adumbrated by Aquinas and its solution already anticipated by this renowned thirteenth century thinker. Yet, temporally situated as he was between Aristotle and Hume, Aquinas does provide an original intermediate viewpoint. While not completely embracing the opposed positions, perhaps his thought can serve to illumine the strengths and weaknesses of both extremes and so reconcile some of the competing claims that continue to be made on behalf of antecedent and simultaneous causality.

The Catholic University of America.

---

[29] For a survey of this literature, see Edward H. Madden, "Scientific Explanations," *The Review of Metaphysics* (1973), pp. 723-43.

THE TITLE *FIRST PHILOSOPHY* ACCORDING TO THOMAS AQUINAS AND HIS DIFFERENT JUSTIFICATIONS FOR THE SAME

JOHN F. WIPPEL

In Q. 5, art. 1 of his *Commentary on the De Trinitate of Boethius*, Thomas Aquinas divides the theoretical sciences on the basis of the different degrees to which objects of theoretical knowledge (*speculabilia*) may be viewed as separated from or joined to matter and motion. He appeals to this criterion in this context to show that the division of speculative science into three parts is fitting.[1] Thus there are certain objects of theoretical knowledge that depend on matter for their very being (*secundum esse*). The proof lies in the fact that they cannot exist except in matter. But these may be subdivided into two classes: 1) Some depend on matter both for their being and for being understood, that is, those whose definition includes sensible matter. Consequently, they cannot be understood without sensible matter and fall under the consideration of physics or natural science. 2) Others, however, even though dependent on matter for their being, do not depend on it for being understood, since sensible matter is not included in their definition. These are the things studied by mathematics.[2]

In addition to the above Aquinas notes that there are other objects of theoretical knowledge that do not depend on matter for their being, since they can exist apart from matter. Some of these are never found in matter, such as God or an angel. Others, such as substance, quality, being, potency, act, the one and the many,

---

[1] *Sancti Thomae de Aquino Expositio super Librum Boethii de Trinitate*, ed. B. Decker (Leiden, 1955), p. 165. Q. 5, art. 1 is directed to this question: "*Utrum sit conveniens divisio qua dividitur speculativa in has tres partes: naturalem, mathematicam et divinam*" (p. 161).

[2] *Op. cit.*, p. 165. Note in particular: "... *quaedam dependent a materia secundum esse et intellectum, sicut illa, in quorum diffinitione ponitur materia sensibilis. ... Quaedam vero sunt, quae quamvis dependeant a materia secundum esse, non tamen secundum intellectum, quia in eorum diffinitionibus non ponitur materia sensibilis, sicut linea et numerus. ...*"

etc., exist in matter in certain cases although not in others. The fact that such objects exist without matter in certain instances suffices to establish Thomas' point here, that they do not depend on matter in order to exist. Such objects, therefore, those that do not depend on matter for their being, will be treated by a third theoretical science, sometimes known as theology or divine science, sometimes as metaphysics, and sometimes as first philosophy.

Thomas comments that this science is entitled theology or divine science because the primary object studied therein is God. He indicates that it is known as metaphysics or as *trans physicam* because it is to be learned by us after physics. In support of this apparent pedagogical reason he appeals to a fundamental tenet of his theory of knowledge, according to which one must proceed from a knowledge of sensibles to knowledge of things that are not sensible. He then notes that it is called first philosophy because all the other sciences, deriving their principles from it, come after it.[3]

Our primary concern in this study is with this third title, first philosophy, and the reasons given here and elsewhere by Thomas in this same work and in his *Commentary on the Metaphysics* for so naming this science. Were one to restrict himself to this passage, the matter would appear to be quite simple. Insofar as other sciences receive their principles from this science, they may be said to come after it. When it is compared with the other sciences, then, it will be called first philosophy because of their dependence upon it. It is interesting to note that in replying to the ninth objection of this same article, Thomas indicates that metaphysics presupposes conclusions proved in the other sciences and also proves the principles of the other sciences. In that immediate context, as supplying principles for the other sciences, he again refers to it as first philosophy and to its practitioner as the first philosopher.[4]

---

[3] *Op. cit.*, pp. 165–66. With respect to the three names, Thomas writes: "*De quibus omnibus est theologia, id est scientia divina, quia praecipuum in ea cognitorum est deus, quae alio nomine dicitur metaphysica, id est trans physicam, quia post physicam discenda occurrit nobis, quibus ex sensibilibus oportet in insensibilia devenire. Dicitur etiam philosophia prima, in quantum aliae omnes scientiae ab ea sua principia accipientes eam consequuntur*" (p. 166).

[4] For an indication of various ways in which other sciences depend on metaphysics according to Aquinas, cf. A. Moreno, "The Nature of

In the *Prooemium* to his *Commentary on the Metaphysics* of Aristotle Thomas again lists the same three titles for this science, theology, metaphysics, and first philosophy. There he has already reasoned that when a number of things are directed to a single goal, one must be director or ruler, and the others directed or ruled. But all the sciences and arts are ordered to one end, the perfection of man, which is happiness. Therefore, one of these sciences should direct or rule the others, and this will deserve the title wisdom. In seeking to determine which science this is, he suggests that it will be the one that is most intellectual. The most intellectual science is that which treats of those things that are most intelligible.[5]

Things may be described as most intelligible from different perspectives, first of all in terms of the order of knowing (*ex ordine intelligendi*). Viewed from this standpoint that science which considers the first causes of things appears to be ruler of the other sciences in fullest measure. This is so because those things from which the intellect derives certitude are more intelligible. Since such are the causes, a knowledge of the causes appears to be most intellectual.

Again, things may be described as intelligible on the basis of a comparison between sense and intellect. While sense knowledge is directed toward particulars intellectual knowledge has to do

---

Metaphysics," *The Thomist* 30 (1966), pp. 132–34. In brief, the particular sciences borrow from metaphysics those concepts that are common to all the sciences such as cause, effect, similitude, substance, accident, relation, essence, existence, etc. They also borrow from it general principles of knowledge such as those of contradiction, identity, etc. It pertains to metaphysics to establish the existence and define the natures of the subjects of the particular sciences, and to defend the principles of such sciences. It should pass judgment on the conclusions of the particular disciplines, rejecting those that are in opposition to its own. It should direct other sciences to its own end. For fuller discussion of Thomas' treatment in his reply to objection 9 of this same Q. 5, art. 1, see our study: "Thomas Aquinas and Avicenna on the Relationship between First Philosophy and the Other Theoretical Sciences: A Note on Thomas' *Commentary on Boethius's De Trinitate*, Q. 5, art. 1, ad 9," *The Thomist* 37 (1973), pp. 133–54.

[5] *In duodecim libros Metaphysicorum Aristotelis expositio*, ed. Cathala-Spiazzi (Turin-Rome, 1950), *Prooemium*. Note in particular: "... *ita scientia debet esse naturaliter aliarum regulatrix, quae maxime intellectualis est. Haec autem est, quae circa maxime intelligibilia versatur.*"

with universals. Given this, that science is most intellectual which treats of the most universal principles, that is, being and those things that follow upon it such as the one and the many, potency and act. Since things of this kind should not remain entirely undetermined and since they should not be treated by any one of the particular sciences, they should rather be examined in one universal science. Being most intellectual, such a science will rule the others.[6]

Finally, something may be described as most intelligible from the standpoint of intellectual knowledge itself. Since a thing has intellective power to the extent that it is free from matter and since intellect and intelligible object must be proportioned to one another, things that are most separate from matter are most intelligible. But those things are most separate or removed from matter which not only abstract from (1) designated matter (*materia signata*) but also (2) from sensible matter altogether, and (3) this not only in the order of thought or definition (*secundum rationem*) but also in the order of being (*secundum esse*). As Thomas had already indicated in Q. 5, art. 1 of his *Commentary on the De Trinitate* and as he again observes here, the first type is studied by natural philosophy, the second is represented by mathematicals, and the third by God and intelligences. Therefore, the science that treats of this last named type of intelligible seems to be most intellectual and chief or mistress of the other sciences.[7]

One might wonder whether or not these different kinds of intelligibles will be investigated by one and the same science. Thomas hastens to make the point that such is indeed the case. The above-mentioned separate substances are universal and primary causes of being (cf. the first and third instances of intelligible objects). Moreover, it belongs to one and the same science to investigate the causes proper to a given genus and to investigate that genus itself. Therefore it will belong to one and the same science to investigate both the separate substances and *ens commune*, for the latter is the "genus" of which the separate substances are the common and universal causes. In other words, while this

---

[6] *Ibid.*
[7] *Ibid.* For his discussion of this in his *Commentary on the De Trinitate* see *op. cit.*, pp. 165–66.

science considers the three aforementioned classes of intelligible objects, it studies only one of them, *ens commune,* as its subject. As Thomas indicates, the subject of a science is that whose causes and properties one investigates rather than the causes themselves of the genus under investigation. Knowledge of the causes of a given genus is rather the end or goal toward which the consideration of the science is directed.[8]

At this point Thomas returns to a distinction among those things that do not depend on matter either for their being or their intelligibility, a distinction we have already seen him making while commenting on the *De Trinitate,* Q. 5, art. 1. As he phrases it here, not only things that are never found in matter such as God and intelligences are said to be separated from matter *secundum esse et rationem,* but also those that may be found without matter, such as *ens commune.*[9]

Then, in the concluding lines of the *Prooemium,* Thomas cites the same three titles for this science that we have already found in his *Commentary on the De Trinitate.* It is to be called divine science or theology insofar as it studies separate substances. It is known as metaphysics insofar as it considers being and that which

---

[8] *Prooemium* to the *Commentary on the Metaphysics: "Ex quo apparet, quod quamvis ista scientia praedicta tria consideret, non tamen considerat quodlibet eorum ut subiectum, sed ipsum solum ens commune. Hoc enim est subiectum in scientia, cuius causas et passiones quaerimus, non autem ipsae causae alicuius generis quaesiti. Nam cognitio causarum alicuius generis, est finis ad quem consideratio scientiae pertingit."* On this also see Q. 5, art. 4 of his *Commentary on the De Trinitate of Boethius,* pp. 190–200. For general discussion of these points see: A. Zimmermann, *Ontologie oder Metaphysik? Die Diskussion über den Gegenstand der Metaphysik im 13. und 14. Jahrhundert* (Leiden-Köln, 1965), pp. 159–80; L. Oeing-Hanhoff, *Ens et Unum Convertuntur: Stellung und Gehalt des Grundsatzes in der Philosophie des. Hl. Thomas von Aquin* (Münster Westf., 1953), pp. 7–20; S. Neumann, *Gegenstand und Methode der theoretischen Wissenschaften nach Thomas von Aquin aufgrund der Expositio super librum Boethii De Trinitate* (Münster Westf., 1965), pp. 113–19; also, J. Doig, "Science première et science universelle dans le 'Commentaire de la métaphysique' de saint Thomas d'Aquin," *Revue philosophique de Louvain* 63 (1965), pp. 41–96; *Aquinas on Metaphysics. A historico-doctrinal study of the Commentary on the Metaphysics* (The Hague, 1972), pp. 55–94; J. Counahan, "The Quest for Metaphysics," *The Thomist* 33 (1969), pp. 519–72.

[9] *In Boet. De Trinitate,* p. 165; *Prooemium, Commentary on the Metaphysics.*

follows upon it. Here Thomas notes that these transphysicals are discovered by the process of analysis (*in via resolutionis*) just as the more universal is discovered after the less universal. It is described as first philosophy insofar as it considers the first causes of things.[10]

Comparison of the reasons offered here in the *Prooemium* for these three titles with those presented in the *Commentary on the De Trinitate* immediately reveals one striking difference. There, as will be recalled, it was described as first philosophy insofar as the other sciences receive their principles from it and come after it. In the *Prooemium* it receives this title because it considers the first causes of things, which first causes Thomas has now identified with the separate substances, that is, God and the intelligences. Moreover, there are other instances in his *Commentary on the Metaphysics* where this second reason is offered to justify entitling this science first philosophy.[11]

Our purpose in the present study, therefore, is this, to attempt to determine more precisely why Thomas sometimes refers to metaphysics as first philosophy because it gives principles to the other sciences, and why he sometimes applies this same name to it because it investigates the first causes. The fact that he does so has been noted by other writers. Why he does so has not, to our knowledge, been satisfactorily determined.[12] It appears to us,

---

[10] "*Dicitur enim scientia divina sive* theologia, *inquantum praedictas substantias considerat.* Metaphysica, *inquantum considerat ens et ea quae consequuntur ipsum. Haec enim transphysica inveniuntur in via resolutionis, sicut magis communia post minus communia. Dicitur autem* prima philosophia, *inquantum primas rerum causas considerat. Sic igitur patet quid sit subjectum huius scientiae, et qualiter se habeat ad alias scientias, et quo nomine nominetur.*"

[11] For other references see J. Doig, "Science première . . . ," p. 43 and n. 4. Cf. in particular *In VI Metaph.* 1, 1170: "*Sed si est aliqua substantia immobilis, ista erit prior substantia naturali; et per consequens philosophia considerans huiusmodi substantiam, erit philosophia prima.*"

[12] Cf. Oeing-Hanhoff, *op. cit.*, p. 17. In commenting on our two passages he writes that this name, first philosophy, "*charakterisiert die Metaphysik nicht nur, insofern sie nach den ersten Ursachen der Dinge fragt, sondern bezeichnet sie auch als die Wissenschaft, die, keine andere voraussetzend, sich selbst begründen muss.*" However, it should be noted that in the reference from the *Commentary on the De Trinitate* (see note 3 above) Thomas does not explicitly state that this science is to be called first philosophy because, in presupposing no other, it must ground itself. Granted

however, that an important clue is given in this same context where Thomas notes that it is called metaphysics insofar as it studies being and those things that follow upon it. He observes that such things that transcend the physical are discovered by the process of analysis just as more universal things are discovered after less universal ones. His reference to analysis at least raises the question as to what process he might be following when he names this science first philosophy because it studies the first causes of things. In any event, we would seem to be well advised at this point to turn to his description of and contrast between analysis and synthesis in Q. 6, art. 1, of that same *Commentary on the De Trinitate of Boethius.*

In the third major part of Q. 6, art. 1, Thomas is attempting to show that one should proceed according to the method of intellect (*versari intellectualiter*) in divine science.[13] In support of this contention he recalls that one may attribute the method of reason (*rationabiliter procedere*) to natural philosophy because this method is most closely followed there.[14] So, too, he will now

---

that this may be a legitimate deduction from Thomas' observation that it is called first philosophy because the other sciences, receiving their principles from it, come after it, it is for this latter reason that it is entitled first philosophy according to Thomas in this context. Moreover, Oeing-Hanhoff leaves unanswered our primary question in the present study: why does Thomas sometimes say that it is called first philosophy insofar as it studies first causes, and why does he sometimes say that it is so named because it gives principles to the other sciences and therefore comes after them? Takatura Ando, in his *Metaphysics: A Critical Survey of its Meaning* (The Hague, 1963), discusses the passages from the *Commentary on the De Trinitate* and the *Prooemium.* Thus on p. 30 he notes that it is called first philosophy "because it deals with the principles which synthetically explain everything that is involved in the special science." But after commenting on the *Prooemium* as well, he does not cite the three reasons offered there by Thomas for the three names, nor does he advert to the different reason offered there for naming it first philosophy. Also, see J. Lotz, "Ontologie und Metaphysik," *Scholastik* 18 (1943), p. 8 and n. 44. While noting the two reasons offered by Thomas to account for the title first philosophy, Lotz does not attempt to resolve our problem. So too Doig, in his *Aquinas on Metaphysics....* Compare p. 69 with pp. 57, 82–83, 86, 91, 173, and 188.

[13] *Op. cit.,* pp. 210–13.

[14] For this, see the first part of this same Q. 6, art. 1, pp. 205–07. Note in particular his concluding remark in the corpus of that part: "*Attribuitur ergo rationabiliter procedere scientiae naturali, non quia ei soli conveniat, sed quia ei praecipue competit*" (p. 207).

maintain that the method of intellect is to be attributed to divine science because it is in this science that this method is most closely followed. After a brief discussion of the difference between reason and intellect,[15] Thomas comments that rational consideration terminates in intellectual consideration according to the process of analysis, but that intellectual consideration is the principle or starting-point (*principium*) of rational consideration according to the process of synthesis or discovery (*secundum viam compositionis vel inventionis*). As he also indicates here, in the process of analysis reason gathers one simple truth from many things. In the process of synthesis, on the other hand, the intellect grasps a multitude of things in one.[16]

Two points have been introduced in this context that are of importance for our purposes: first, the distinction between the process of analysis and that of synthesis; second, the observation that intellectual consideration may be regarded as the terminus of rational consideration according to the process of analysis, while it may be regarded as the beginning or starting-point of rational

---

[15] *Op. cit.*, p. 211. For fuller discussion of the relationship and distinction between *intellectus* and *ratio* in Thomas see J. Peghaire, *Intellectus et Ratio selon S. Thomas d'Aquin* (Paris-Ottawa, 1936).

[16] "*Sic ergo patet quod rationalis consideratio ad intellectualem terminatur secundum viam resolutionis, in quantum ratio ex multis colligit unam et simplicem veritatem. Et rursum intellectualis consideratio est principium rationalis secundum viam compositionis vel inventionis, in quantum intellectus in uno multitudinem comprehendit*" (*op. cit.*, p. 211). If Thomas here associates *ratio* with analysis and *intellectus* with synthesis, this follows from an analogy which he has just drawn between multitude and unity, on the one hand, and analysis and synthesis, on the other. Thus: "*Differt autem ratio ab intellectu, sicut multitudo ab unitate . . . Est enim rationis proprium circa multa diffundi et ex eis unam simplicem cognitionem colligere . . . Intellectus autem e converso per prius unam et simplicem veritatem considerat et in illa totius multitudinis cognitionem capit . . .*" (*ibid.*). In brief, then, reason, which gathers one simple truth by starting from a many, is characteristic of the process of analysis; whereas, intellect, which first contemplates one truth and in that truth grasps a many, is more characteristic of the process of synthesis. For a general discussion of the distinction between analysis and synthesis in Thomas see L.-M. Régis, "Analyse et synthèse dans l'oeuvre de saint Thomas," *Studia Mediaevalia in honorem admodum reverendi patris Raymundi Josephi Martin* (Bruges, 1948), pp. 303–30; also, S. Edmund Dolan, "Resolution and Composition in Speculative and Practical Discourse," *Laval théologique et philosophique* 6 (1950), pp. 9–62; Doig, *Aquinas on Metaphysics . . .* , pp. 64–76.

FIRST PHILOSOPHY

consideration according to the process of synthesis. Since Thomas is attempting to establish the point that the method of intellect is to be attributed to divine science, he then comments that the kind of consideration which is the terminus of all human reasoning is intellectual to the maximum degree. According to the process of analysis, all rational consideration in all the sciences terminates in the consideration of divine science.[17] In support of this he now introduces a further distinction. As he has already indicated above,[18] reason sometimes moves from knowledge of one thing to knowledge of another in the order of reality (*secundum rem*), as when there is demonstration through extrinsic causes or effects. Reason may thus advance either according to the process of synthesis, by moving from cause to effect, or according to the process of analysis, by moving from effect to cause. This application of synthesis and analysis to cause and effect reasoning can be made because causes are simpler, more unchangeable, and more constant than their effects, and because, as has been observed above, in the process of analysis one gathers one simple truth from a many.[19] Hence in this case, by the process of analysis one arrives at knowledge of that which is simpler (cause) by moving from knowledge of that which is less simple (the effect). Consequently, the ultimate term of the process of analysis when one reasons

---

[17] *Op. cit.*, pp. 211–12. "*Illa ergo consideratio, quae est terminus totius humanae ratiocinationis, maxime est intellectualis consideratio. Tota autem consideratio rationis resolventis in omnibus scientiis ad considerationem divinae scientiae terminatur.*"

[18] Q. VI, art. 1, first part (*op. cit.*, pp. 206–07): "*Secundo, quia cum rationis sit de uno in aliud discurrere, hoc maxime in scientia naturali observatur, ubi ex cognitione unius rei in cognitionem alterius devenitur, sicut ex cognitione effectus in cognitionem causae. Et non solum proceditur ab uno in aliud secundum rationem, quod non est aliud secundum rem, sicut si ab animali procedatur ad hominem. In scientiis enim mathematicis proceditur per ea tantum, quae sunt de essentia rei, cum demonstrent solum per causam formalem . . . Sed in scientia naturali, in qua fit demonstratio per causas extrinsecas, probatur aliquid de una re per aliam rem omnino extrinsecam.*"

[19] *Op. cit.*, p. 212. "*Ratio enim, ut prius dictum est, procedit quandoque de uno in aliud secundum rem, ut quando est demonstratio per causas vel effectus extrinsecos: componendo quidem, cum proceditur a causis ad effectus; quasi resolvendo, cum proceditur ab effectibus ad causas, eo quod causae sunt effectibus simpliciores et magis immobiliter et uniformiter permanentes.*"

secundum rem or in terms of extrinsic causes is a knowledge of supremely simple causes, that is, the separate substances.[20]

At other times, however, reason moves from one thing to another in the order of reason (*secundum rationem*), as when one proceeds in terms of intrinsic causes. This too can occur either according to the process of synthesis or according to the process of analysis. In the first case one advances from the most universal forms to the more particular. In the second case one proceeds in reverse order, that is, from the more particular to the most universal. This is so, suggests Thomas, because what is more universal is simpler and, as has already been noted, in its process of analysis reason gathers one simple truth from a many, here the most universal from the more particular.[21] But, continues Thomas, the most universal things are those that are common to all beings. Therefore, the ultimate terminus of analysis in this movement of reason in terms of intrinsic causes (*secundum rationem*) is a consideration of being and that which pertains to being as such.[22]

In sum, therefore, Thomas has now shown that the ultimate term of the process of analysis when one reasons in terms of extrinsic causes (*secundum rem*) is a knowledge of separate substances. Its ultimate term when one proceeds according to intrinsic causes (*secundum rationem*) is a knowledge of being and the properties that follow from being as such. But as he has already shown above in Q. 5, art. 1 and Q. 5, art. 4, these two, being and separate substances, are the things of which divine science treats.[23] Moreover, since he has already indicated in the present context that that consideration is intellectual to the maximum degree which is the terminus of all human reasoning, we are not

---

[20] *Ibid.* "*Ultimus ergo terminus resolutionis in hac via est, cum pervenitur ad causas supremas maxime simplices, quae sunt substantiae separatae.*"

[21] *Ibid.* "*Quandoque vero procedit de uno in aliud secundum rationem, ut quando est processus secundum causas intrinsecas: componendo quidem, quando a formis maxime universalibus in magis particularia proceditur; resolvendo autem quando e converso, eo quod universalius est simplicius.*"

[22] *Ibid.* "*Et ideo terminus resolutionis in hac via ultimus est consideratio entis et eorum quae sunt entis in quantum huiusmodi.*"

[23] *Op. cit.*, pp. 165–66; 195.

surprised to find him concluding that the consideration of divine science is supremely intellectual.[24]

At this point in the discussion Thomas again briefly considers the different names assigned to divine science. The name divine science itself is not at issue here, since this is the title used by Boethius in the text upon which Thomas is commenting, and primarily by Thomas himself until this point in Q. 6, art. 1.[25] Because intellectual consideration is the principle or starting-point of rational consideration, Thomas now comments that divine science gives principles to all the other sciences. For this reason it is called *first philosophy*. But because intellectual consideration may also be regarded as the terminus of rational consideration, this science is learned after physics and the other sciences. For this reason it is called *metaphysics,* beyond physics, as it were, because it comes after physics according to the process of analysis.[26]

With respect to the title first philosophy, therefore, it will be recalled that Thomas holds that intellectual consideration is the principle of rational consideration according to the process of synthesis. If he now styles this science first philosophy because it gives principles to the other sciences, this is because intellectual consideration is the principle of rational consideration. But according to what we have now seen, this in turn is true only according to the process of synthesis, not according to the process of analysis. Consequently, when Thomas names this science first philosophy because it gives principles to the other sciences, we

---

[24] *Op. cit.*, p. 212.

[25] Note that Q. VI, art. 1 is directed to the following question: "*Utrum oporteat versari in naturalibus rationabiliter, in mathematicis disciplinabiliter, in divinis intellectualiter*" (*op. cit.,* p. 201). As we have already indicated above, in the third part of this question Thomas attempts to show that one should proceed according to the method of intellect in *divine science* (*op. cit.,* pp. 210ff.). For the text of Boethius see his *De Trinitate,* 2 (*Boethius: The Theological Tractates with an English Translation,* ed. H. F. Stewart and E. K. Rand [Cambridge, Mass., 1968]), p. 8: ". . . *in naturalibus igitur rationabiliter, in mathematicis disciplinaliter, in divinis intellectualiter versari oportebit . . .*"

[26] *Op. cit.*, p. 212. "*Et exinde etiam est quod ipsa largitur principia omnibus aliis scientiis, in quantum intellectualis consideratio est principium rationalis, propter quod dicitur prima philosophia; et nihilominus ipsa addiscitur post physicam et ceteras scientias, in quantum consideratio intellectualis est terminus rationalis, propter quod dicitur metaphysica quasi trans physicam, quia post physicam resolvendo occurrit.*"

conclude that he is then regarding its consideration as the principle or starting-point of the other sciences, to be sure, but that when he so correlates this science and others he is considering the movement from one to the other according to the process of synthesis.

Moreover, in the preceding context he has distinguished between the advance of reason in the order of intrinsic causes (*secundum rationem*) and its advance according to extrinsic causes (*secundum rem*). He has noted that the processes of synthesis and analysis may be applied to either of these. The ultimate terminus of analysis in the movement of reason according to intrinsic causes was identified with being and the properties that belong to being as such. Consequently, it would seem to follow that when Thomas entitles this science first philosophy because it gives principles to the other sciences, he is then viewing the movement from this science to the others according to the process of synthesis rather than according to the process of analysis, as we have already suggested, and in the order of intrinsic causes (*secundum rationem*) rather than in the order of extrinsic causes (*secundum rem*). In other words, he is viewing it as the science of being as being and suggesting that it gives principles to the other sciences insofar as more particular concepts and more particular principles follow from the most general concepts (being, etc.) and most general principles. As he indicates in other contexts, just as it pertains to this science to study being as being, so too does it pertain to it to examine and defend the principles that follow immediately from being as being.[27] Granted this is a fuller explanation of his reasoning for so entitling this science than one finds explicitly stated in the text itself, it seems to be a logical deduction from distinctions that he has made there, and also appears to be consistent with the view expressed in Q. 5, art. 1.

---

[27] On this see the remarks and texts found in Moreno's study, "The Nature of Metaphysics," cited in n. 4 above. Also cf. J. Counahan, *op. cit.*, pp. 531–33; Oeing-Hanhoff, *op. cit.*, p. 18. Note the following from the *Summa contra gentiles* III, c. 25: "*Hoc autem modo se habet philosophia prima ad alias scientias speculativas, nam ab ipsa omnes aliae dependent, utpote ab ipsa accipientes sua principia, et directionem contra negantes principia: ipsaque prima philosophia tota ordinatur ad Dei cognitionem sicut ad ultimum finem, unde et scientia divina nominatur.*"

FIRST PHILOSOPHY

There, it will be recalled, this science is named first philosophy because the other sciences, receiving their principles from it, come after it.[28]

As we have seen above, in the *Prooemium* to his *Commentary on the Metaphysics* Thomas offers a different reason for describing this science as first philosophy. There he writes that it is so named because it considers the first causes of things. In that context it is clear that the latter are to be understood as extrinsic causes.[29] Hence, if we may apply the distinction between the movement of reason in terms of intrinsic causes (*secundum rationem*) and in terms of extrinsic causes (*secundum rem*) of Q. 6, art. 1 of his *Commentary on the De Trinitate* to the present context, the *Prooemium*, it is reasoning in terms of extrinsic causes (*secundum rem*) that Thomas now has in mind. Moreover, if we wonder whether the process of analysis or that of synthesis is at issue here, it appears to be the former rather than the latter. As Thomas has already stated in Q. 6, art. 1, the ultimate term of the process of analysis when one reasons in terms of extrinsic causes is attained when one arrives at a knowledge of supremely simple causes, that is to say, the separate substances.[30] Consequently, when metaphysics is described as first philosophy because it gives principles to the other sciences, it is being viewed in terms of intrinsic causes (*secundum rationem*) and according to the process of synthesis with respect to these sciences. Such is the standpoint of Q. 5, art. 1 and Q. 6, art. 1 of the *Commentary on the De Trinitate*. But when it is so named because it studies the first causes, it is being viewed in terms of extrinsic causes (*secundum rem*) and apparently according to the process of analysis with respect to these causes, since according to this perspective one reasons from a knowledge of effects to a knowledge of separate substances as their causes. Such is the standpoint implied by Thomas' explanation of the reasons for this title in the *Prooemium* to the *Commentary on the Metaphysics*.

Two further possibilities present themselves at this juncture. Why not view this science in terms of extrinsic causes (*secundum*

---

[28] For the text see note 3 above.
[29] For the text see note 10 above.
[30] See note 20 above.

rem), but according to the process of synthesis? Or why not view it in terms of intrinsic causes (*secundum rationem*), but according to the process of analysis? According to the first suggestion one would then move in this science from a knowledge of God and separate substances to a knowledge of effects that follow from the same. While such might indeed obtain in a universe wherein man enjoys some kind of direct insight into the divine essence and wherein there is no freedom on the part of God to create or not to create, neither of these conditions would be conceded by Thomas Aquinas. Hence there would have been little point in his viewing metaphysics in terms of extrinsic causes according to the process of synthesis.

As to the second suggestion, considering metaphysics in terms of intrinsic causes (*secundum rationem*) and according to the process of analysis with respect to its relationship to the other sciences, Thomas does concede this possibility, but suggests that it should be called metaphysics (rather than first philosophy) when it is so viewed. Thus in Q. 6, art. 1 he notes that it is learned after physics and the other sciences because intellectual consideration is the terminus of rational consideration. As he has explained in this same context, when one advances according to intrinsic causes (*secundum rationem*) by the process of analysis, one proceeds from the more particular to the more universal. Since that which is most universal is common to all beings, the ultimate term of analysis *secundum rationem* is being and that which pertains to being as such. To return again to his discussion of the title, metaphysics, he then comments that it is for this reason that it is called metaphysics or *trans physicam,* because it comes after physics according to the way of resolution (analysis).[31]

This simply reinforces the view expressed in Q. 5, art. 1, according to which it is called metaphysics or *trans physicam* because it is to be learned by us after physics. There he has noted that we must move from a knowledge of sensible things to a knowledge of things that are not sensible.[32] In that immediate context he does not explicitly indicate whether these not-sensible things include the

---

[31] See note 26 above.
[32] See note 3 above.

FIRST PHILOSOPHY 599

two classes of things that do not depend on matter for their being and to be defined, that is, those that are never found in matter such as God and the angels, and those that are sometimes found in matter and sometimes not, such as substance, quality, being, etc. He simply notes that this science treats of them all and that it is known as theology, as metaphysics, and as first philosophy, as we have already seen. Perhaps, then, in Q. 6, art. 1, one might extend his reason for naming it metaphysics to both classes, and suggest that it is so named or comes after physics in the order of resolution (analysis) both as applied in terms of intrinsic causes (*secundum rationem*), hence to being and its properties, and in terms of extrinsic causes (*secundum rem*), hence to God and separate entities. However, in the light of the discussion of the title metaphysics in the *Prooemium,* we are more inclined to restrict his reason for so entitling it in Q. 5, art. 1 and Q. 6, art. 1 to the first of these. Thus in the *Prooemium* he notes that it is called metaphysics insofar as it considers being and those things that follow upon being. For *these* transphysicals are discovered by way of analysis (*in via resolutionis*) just as the more general is discovered after the less general.[33] Here it is clear that he is appealing to the process of analysis and to the movement of reason according to intrinsic causes (*secundum rationem*) alone when it comes to justifying the name metaphysics. It would seem more likely, then, that such was also his intention in the two discussions in the *Commentary on the De Trinitate.*

In conclusion, therefore, when Thomas styles this science metaphysics whether in the *Commentary on the De Trinitate* or in the *Prooemium* to his *Commentary on the Metaphysics,* it is because it comes after physics in the order of analysis. This is so because it studies being and its properties and because these are discovered after one has investigated sensible things, at least to some extent, and because one is dealing with a movement of reason according to intrinsic causes (*secundum rationem*) rather than according to extrinsic causes (*secundum rem*). When he names it first philosophy because it gives principles to other sciences, this is because he is viewing the movement of reason from

---

[33] See note 10 above.

it to other sciences in terms of intrinsic causes (*secundum rationem*) once again, to be sure, but according to the process of synthesis rather than the process of analysis. When he names it first philosophy because it studies the first causes of things he is now viewing the movement of the mind according to extrinsic causes (*secundum rem*), and according to the process of analysis or the way of resolution rather than according to the process of synthesis.

The Catholic University of America.

BOOKS RECEIVED

SUMMARIES AND COMMENTS *
WILLIAM A. FRANK AND STAFF

BARTLEY, W. W., III. *Wittgenstein.* Philadelphia: J. B. Lippincott Company, 1973. 192 pp. $6.95—This book is an intellectual biography of Ludwig Wittgenstein covering the decade following the First World War. For the most part the work is narrated after the fashion of a field research journal and is filled with incidents and anecdotes that are new to Wittgenstein lore. The book has three major sections. The first discloses previously unrevealed aspects of Wittgenstein's character and personal life with the open shamelessness common to contemporary writers. The second part is devoted to a consideration of *"Der Satz,"* Wittgenstein's preferred title for the *Tractatus*. Bartley claims that the main thrust of the work was a theme common to the major turn-of-the-century German and Austrian writers, Hofmansthal, Karl Kraus, Kafka, and Mauthner, namely, that "speech was becoming a vehicle not of communication but of mendacity and pointlessness" (59). Bartley hopes to show how Wittgenstein's "significant silence" is effected not by forging a classic of epistemological empiricism nor by writing a Kantian treatise but rather through an explication of the pre-Kantian belief that somehow language mirrors the world.

With this orientation Bartley conjectures in the third section that the shift from the *Tractatus* was primarily influenced by Karl Bühler, whom he, Wittgenstein, studied while preparing for those teaching duties that occupied him during the decade under study. Bühler is shown to concentrate upon the developmental aspects of child psychology, espousing a theory of "imageless intentionality" wherein the mind organizes independently of any particular primitive element. For Wittgenstein, the claims and nature of philosophy were to rest on language analysis, and this analysis was not unconnected to his at-

---

* Books received are acknowledged in this section by a brief resume, report, or criticism. Such acknowledgement does not preclude a more detailed examination in a subsequent Critical Study. The Summaries and Comments are written by the Manager and his Staff (listed on the masthead), with the help of others. In addition to the staff, B. M. Bonansea, Frederick E. Brenk, Charles F. Breslin, Charles A. Corr, Donald A. Cress, George Dalin, DuWayne Engelhardt, William Gerber, Adrian Gilligan, Milton Goldinger, William J. Hill, Burton G. Hurdle, Jr., Justin Leiber, Michael P. Malloy, Camille Minichino, Robert L. Perkins, John Rudoff, Wilhelm A. Schmidt, Robert Sokolowski, Leo Sweeney, and Jeff White have contributed to this issue. From time to time, technical books dealing with such fields as mathematics, physics, anthropology, and the social sciences will be reviewed in this section, if it is thought that they might be of special interest to philosophers.

tempt to make room for responsible human action. The core of Bartley's work is an effort to understand the relationship between Wittgenstein's written philosophy and his ultimate concerns. Bartley's researches into the life of the compassionate, original, awesome Austrian school teacher make for good reading and, more importantly, suggest an as yet obscure significance to the man's thought.—W. A. F.

BEVAN, R. A. *Marx and Burke: A Revisionist View.* La Salle: Open Court Publishers, 1973. 197 pp. $7.95—This book ranges over a number of problems in contemporary political science. Ostensibly about Marx and Burke, Aristotle and the behaviorists also figure in the development. One of the major difficulties of the book is the forced presence of Aristotle and the absence of Hegel. "Burke and Marx being in the Aristotelian tradition, considered it absurd to speak of man as anything but a social or political animal—a *zoon politikon* . . ." (p. 61). This sentence is probably correct, but it is *ad hoc,* for, on this point, Hegel is Marx's teacher—not Aristotle. There are numerous such references, and thus, as a revisionist view in the history of ideas, the book seems at points strained and at others inaccurate. There is a level of generality at which one can say that Burke, Aristotle, and Marx occupy a common position, and that is the level at which this book is written, viz., a very *general* generality. One is simply stunned by the overcoming of differences, the omission of distinctions. To suggest that Hegel is a Rationalist (capital "R" there) is to stretch the language of philosophy to meaninglessness (p. 21). Repeated again is the simplicities about thesis, antithesis, synthesis, cloppity, clop (p. 22). Perhaps this is enough about the shortcomings of this book.

The very *general* generality which has been my complaint constitutes the revisionist view. All too often by making distinctions we overlook the common ground, and Bevan has again revealed the common ground of most of western political theory. The *Grundlagen* are common and they should be recalled now and again. The *Grundlagen* underlie both Burke and Marx and unite them against the behaviorists. It is here that the contemporary debate is joined. To be sure, one could approach a critique of behaviorial political science from specifics, say, from Burke's *versatile ingenium* and the Marxist notion of "bourgeois science." The recall of the commonplaces is worthwhile, however, because of the long-standing testimony they constitute. The historical tradition and the commonplaces do not constitute a mere argument from authority, as the modern ahistoricalist would insist: an argument is not false because it is old or first spoken by a person now deceased. Rather, as Bevan argues, it is the rejection of history, the tradition and its insights that constitutes the behaviorial contribution, and this rejection applies to Burke and Marx alike equally. There is no confusion regarding Bevan's position in the contemporary debate.—R. L. P.

BLACKBURN, S. *Reason and Prediction.* New York: Cambridge University Press, 1973. 175 pp. $11.50—Blackburn's book is concerned with a "certain way of reasoning" corresponding to that "belief which we have because we expect uniformities in our experience to be representative." In short, the subject is the problem of induction and the purpose is the redoubtable task of justifying inductive reasoning. The fundamental question to which the author addresses himself is "whether, and why, we are ever right to use I-evidence as a reason for an I-conclusion?" (where I-evidence is the statement that is used as a reason for some other statement and I-conclusion is the statement for which we are reasoning). While Blackburn presents a sustained and intricate argument, perhaps four of the major aspects of his study deserve particular mention. First, as a result of an inquiry into the concept of a reason, it becomes apparent that some sort of normative analysis of the concept is indicated. For one criterion of a reasoning procedure must be reliability; such a procedure should generally result in the truth. Moreover, there is a distinct relationship between having confidence in a proposition and that proposition being true. Ultimately, "the authority of reasoning is nothing more, nor less, than the authority of truth."

Second, in considering the nature of the similarities and resemblances for which we reason, Blackburn employs Goodman's paradox as something of a stalking-horse. In controverting Goodman's own solution to the paradox, it is argued that there is, in fact, a serious logical asymmetry between "grue" and "blue" such that "to tell that something is grue *entails* telling not only what it looks like but also what time it is, or what time it was first examined, whereas telling that something is blue does not entail either of these things." Third, it is Blackburn's contention that a justification of inductive reasoning can and must be provided without recourse to the theory of probability not only because of the theory's inconsistent internal aspects (the objective vs the subjective aspects) but also because of the foothold which it can provide for the vigilant and ever present skeptic. Finally, in the last three chapters of the book, Blackburn defends inductive reasoning by demonstrating its connection with success in prediction. In the end, the author's conclusion is both cautious and conservative: rather than claiming that we can have confidence in statements "covering all objects of all sorts under all conditions at all times," we can at most certify success (in predicting) for "fractional extrapolations of regularities." Although Blackburn's presentation is complex and somewhat technical, this is to be expected. For while the problem can be stated in deceptively simple terms, the solution is considerably less elementary. Certainly, the book offers a persuasive alternative to various more skeptical views of inductive reasoning which have appeared in recent years and contains numerous principles and concepts which are worthy of close attention.—B. G. H.

BOGEN, J. *Wittgenstein's Philosophy of Language.* New York: Humanities Press, Inc., 1972. 244 pp. $10.00—The book is published in the International Library of Philosophy and Scientific Method. It proceeds under the assumption that the *Tractatus* and the later works of Wittgenstein are mutually illuminating. The general program is to present the Tractarian picture theory, to explain why it was abandoned and a new theory of language adopted, and to explicate the new theory of use. Conceptually the book is arranged around the problem of intentionality. Bogen believes that Wittgenstein's chief concern was with the problem of false belief.

Perhaps the best part of the book is the first chapter where the author sketches his view of the Tractarian ontology. Bogen presents its presuppositions, its major claims, and its internal problem. Picture theory of language presupposes that facts are real and individual, and that the world has a fixed structure. It claims that a proposition *represents* reality, and that the proposition, as an assertion, *presents* a possible fact. Hence a proposition is also a fact. If facts are individuals and exist independently of one another, the internal problem is to show how assertion and fact can be essentially connected. Bogen's second chapter explains how this problem forced Wittgenstein to abandon the picture theory as an adequate account of language. Instead, Wittgenstein turned to a *use* theory of language in which facticity loses its ontological independence and is tied very closely to asserting.

Bogen observes that this shift in Wittgenstein moves him from an explanation to a description of language, from the discovery of the hidden to account for language, to clarifying what is already manifest in language. Hence the last chapter proposes to explicate *use* as a calculus, which in its pure, rationalistic form proves unhelpful. Bogen then thinks that to explain *use*, Wittgenstein was led to cast the meaning of language behavior into a total web of behavior patterns, life-forms, and history. Regarding picture theory, the author is extremely helpful. He is interesting but not convincing on the later Wittgenstein. Too much is said in the *Investigations* which is not illuminated by Bogen's study, and what does one do with the very last parts of the *Tractatus!* Although the problem of intentionality is central to much contemporary philosophy and even though it was perhaps directly inspired by Wittgenstein's investigations, one can wonder whether Wittgenstein saw it as his own central problem.—W. A. F.

CARTER, C. L. (ed.). *Skepticism and Moral Principles.* Evanston, Illinois: New University Press, Inc., 1973. 145 pp. $5.95—This volume, designed to bring together new analyses of moral skepticism, consists of papers by Professors William Frankena, Marcus Singer and Antony Flew and a long introduction by the editor which describes the central issues and discusses each of the papers. In his paper, "The Principles of Morality," Frankena contends that underlying many of our ordinary moral expressions is the implicit belief in an absolute moral action-guide, i.e., an action-guide which all those who are fully rational within the moral point of view would acknowledge. Although Frankena does not discuss in this paper the crucial and perplexing problems of whether all those taking the moral point of view must accept the same principles and whether there is an irreducible plurality of moral points of view, he does discuss at length H. D. Aiken's rejection of moral absolutism in favor of moral autonomy and the differences between his theory and R. Firth's Ideal Observer Theory. In "Moral Skepticism," Singer carefully defines his topic and classifies its major variants, three of which—personal subjectivism, social subjectivism and emotivism—he criticizes in detail. Although Singer does not present new arguments against the various forms of moral skepticism, he provides the valuable service of organizing clearly and stating precisely the major difficulties. The concluding paper by Flew, "Must Morality Pay? or What Socrates Should Have Said to Thrasymachus," seems out of place in this collection since it neither defends an absolute position nor attacks a variety of skepticism. Flew argues that Socrates' attempt to prove that the life of justice is always in one's interest is misguided, since, in the real world, morality often demands that an individual sacrifice his good for the good of others. What Socrates should have argued, according to Flew, is that the choice of the moral life rather than the prudent one cannot be deemed to be either rational or irrational because for such a choice, the issue of rationality arises only in connection with the coherence of our plans and purposes or the internal consistency of our preferences.—M. G.

CHOMSKY, N. *Studies on Semantics in Generative Grammar*. The Hague: Mouton, 1972. 207 pp. G 27—Three long papers are collected here which constitute Chomsky's major theoretical work on syntax and semantics subsequent to the "standard theory" of *Aspects of the Theory of Syntax* (1965). Since 1965, transformational-generative linguists have suggested various changes in "standard theory," centering on the relationship between the syntactic and semantic components in natural language grammars. In these papers Chomsky explains several specific problems that require the extension of standard theory and he criticizes the proposals and arguments of the generative semanticists, attempting to show that many of the specific features of language that are claimed to support the generativist position are as well or better explained by extended standard theory, and emphasizing that most of the generativist's proposals turn out to be mere "notational variants," evidentially indistinguishable from extended standard theory. Aside from appealing to particular data favoring his "lexicalist" defense of syntactic deep structure, Chomsky makes a Popperean appeal that his position is preferable because it is more constraining and more falsifiable.

In the first of these papers, "Remarks on Nominalization," Chomsky considers one specific sort of evidence that he thinks favors extended standard theory over the generative semanticists. He there contrasts "gerundive nominals" such as "John's being eager to please" with "derived nominals" such as "John's eagerness to please." The point is that gerundives are sentence-like and have a clear and regular transformational relationship with sentences that they nominalize (e.g., "John is eager to please"), but this is not the case with the derived nominals: extended standard theory easily explains this as the difference between a transformational and syntactic relationship, in the case of gerundives, and a lexical and semantic relationship, in the case of the derived nominals. Gerundives have the same syntactic deep structure as what they nominalize, while derived nominals have a different deep structure. But this form of explanation is not available to the generative semanticist in that he denies that syntactic deep structure is well-defined and distinguishable from the full logico-semantic specification of a sentence. "Deep Structure, Surface Structure, and Semantic Interpretation" and "Some Empirical Issues in the Theory of Transformational Grammar" range over a number of similar sorts of arguments in favor of the interpretivist position: in these two papers the need for reducing the power of grammatical rules, and increasing the falsifiability of their theory, is prominent. It is evident that transformational-generative grammatical theory is in a state of flux and turmoil.—J. L.

DESCARTES, R. *Treatise of Man.* (French text with translation and commentary by Thomas Steele Hall.) Cambridge: Harvard University Press, 1972. 232 pp. $11.00—Sixth in the Harvard Monographs in the History of Science Series, this volume is both an excellent addition to the history of science and to Cartesian studies. Taking E. Gilson's *Index scholastico-cartésien* for his inspiration, Hall offers us, by way of an excellent and well-documented set of annotations to his translation, an interesting view of the place of Descartes' work in the history of science. Contained in the volume are: a foreword by I. Bernard Cohen (general editor of the Harvard Series); a list of abbreviations; an introduction; a synopsis of the physiology of Descartes; a synopsis of the contents of the first French edition of 1664; a list of bibliographical materials containing editions of the *Treatise of Man* and secondary sources "for the study of Descartes' physiology;" the English translation and commentary; a facsimile of the first French edition; an index.

Hall makes a number of cross-references to the *Description of the Body,* pointing out differences between doctrines found in this work and the *Treatise of Man.* Although it is easy to mistake the first French edition for the Adam-Tannery edition, the text reproduced is not from the Adam-Tannery edition, but rather from the 1664 edition of Clerselier. Nevertheless, Hall does supply in his English translation the corresponding Clerselier and Adam-Tannery page numbers, which makes looking up a translated text quite simple. The book is an altogether pleasant addition to Cartesian scholarship. Much of the attempts at determining Descartes' sources have been aimed at determining his philosophical sources. It is time that Descartes' medical and physiological sources have been uncovered for us. The notes alone should recommend this text. They are a gold mine of references to 16th and 17th century medical history.—C. A. C.

DESAN, W. *The Planetary Man.* New York: The Macmillan Co., 1972. 380 pp. $9.95—This is a highly lucid, sensitive study of the problem of human survival. Desan began his analysis with a noetic critique of man in the social order. In the second volume, published now for the first time, he extends this study into the ethical implications of man's search for a united world. In the context of our own crisis-ridden times, Desan sees a need for a new approach to the problem of human survival. He provides a methodological structure that acts as a corrective to the almost Cartesian egocentricity of Existentialism. A new vocabulary is provided to afford a tolerant recognition to the Other in the social context. Desan feels that Existentialism has not offered an entirely adequate analysis of larger social phenomena, since these were beyond the individual. Desan does employ a phenomenological method, but it is a "descending" one rather than the usual "ascending" method. Instead of analyzing in terms of the individual *outward,* as in the work of Husserl or Sartre, the "descending" method begins with the larger social reality and works *inward.*

This analysis reveals to the philosopher-Observer a social reality appearing as a *totum physicum humanum,* a vast, complex continuum of human individuals. They are interwoven into an ever-changing pattern whose very existence is sustained by the interdependence of the human components. Human toleration and consideration is seen as more than an ethical postulate; it is a fundamental prerequisite for survival. The *totum* is not entered into by an act of the individual's will. As we live, we are part of it. We are sustained, we grow and learn, because the social whole nourishes us in a multitude of ways. There is no isolation from the *totum;* individual existence requires the sustenance of the social context. In this sense, we are each seen as a "fragment" of the *totum,* an intersection of various interdependent space-time events. The fragment is only aware of a limited portion of this interwoven *totum* because of his particularized position within it. This finitude is called "angularity" by the author. It is an inherent limitation, noetically and ethically, of the fragment vis-a-vis the *totum*. As we reflect on this existence, and as we become more aware of the phenomena surrounding and nurturing it, we are able to widen the scope of that angularity. The ideal towards which the philosopher strives is that of the "Planetary Man." It is he who attempts a noetic integration of all within the *totum* consciously and completely. In recognizing our own failure to reach that unity, and in thus acknowledging our inherent angularity, we begin to recognize the need for a sense of tolerance towards the other fragments.

To act with tolerance, to take account of the fact of each fragment's angularity, is the duty of the individual. It is his responsibility not only to the other individuals, but also to the *totum* as a whole. The social whole can impose positive obligations upon him to further the aim of group survival. In this context, the ideal of the "Planetary Man" becomes the "mundane Saint," who tries to integrate man's action as well as his knowledge. These ethical considerations have wide implications. In a sense, Desan's method is a descriptive metaphysics. The ontological relationship between the *totum* and its fragments is a subtle one, and it will doubtless prompt much further investigation. Clearly, Desan claims for the *totum* an independent status that will put him at odds with Existentialism on another front. Seen as a corrective, *The Planetary Man* gives greater consideration to the contribution of the social whole to the development of the individual entity. Viewed independently, this book is a fascinating critical analysis of man and his intersubjective world.—M. P. M.

ELDERS, L. *Aristotle's Theology. A Commentary on Book XII of the Metaphysics.* New York: Humanities Press, 1972. 309 pp. $18.50
—This is a careful, line-by-line and often word-by-word commentary on Book XII of the Metaphysics. The commentary is preceded by a seven part introduction which deals with the theology of Book XII, *noûs,* self-knowledge, desire, the place of the book in Aristotle's writings, its date and structure, and the problem of Chapter 8 and Aristotle's monotheism. Elders claims Chapter 8 was not written by Aristotle but by a disciple or disciples. He also claims that Book XII contains at least five other distinct treatises which come from different periods in Aristotle's life. Throughout his book Elders summarizes the opinions of all the important modern and ancient commentators who have written on the questions he examines, and makes copious references to other Greek thinkers and other works of Aristotle. For example the section on self-knowledge moves through several dialogues of Plato and through Aristotle's ethical writings. Philological observations abound, and Elders is sensitive to philosophical aspects in them. Some of his remarks about terms like *ousia* and *dokei* contain helpful philosophical insights. The presentation is lean, clear and direct. Elders has marked off another definite part of Aristotle's *Metaphysics* (his earlier work was on Book X) and has supplied us

with all the information, sources and scholarly commentary that are available for it.—R. S.

FACULTY IN PHILOSOPHY, American University. *Explanation: New Directions in Philosophy.* Edited by Barry L. Blose and others. The Hague: Martinus Nijhoff, 1973. 216 pp. G 33.80—Can contemporary American philosophy be characterized in a way which would meaningfully distinguish it from philosophy in other times, other places? The somewhat negative answer to this question given by John E. Smith (*The Spirit of American Philosophy*, 1963) and Andrew J. Reck (*The New American Philosophers*, 1968) is a source of puzzlement to Roger T. Simonds, author of the first essay in this collaboration volume. Simonds asserts, with reference to activism, pragmatism, and optimism, that while "these qualities are . . . not the exclusive property of American philosophers," yet "Americans seem to show them more than others do," and this is a meaningful characterization because "America herself is a sort of incarnation of Western activism, pragmatism, and optimism." Whether Western philosophy as a whole can profitably take a new point of departure, viz., that of Shankara, the ninth century systematizer of Vedantism, is the question dealt with in the second essay of this volume, written by Cornelia D. Church. Both Shankara and Western philosophers attempt to bring into focus God (the unconditioned), the world (reality), man (the good), and mind (knowledge, truth), but Shankara tries to see how man can be liberated from his limits of perception through a quasi-mystical technique. The possibilities of this avenue for Western seekers of truth, according to Miss Church, "are tremendously exciting."

The remaining eight essays concern the relation between explanation, on the one hand, and behavior, language, religion, and philosophy, on the other. A brilliant and uncommonly absorbing essay by Barry L. Blose subjects to inquisitorial scrutiny a famous objection against the view that physical-thing language and sense-datum language are intertranslatable. The objection points out that a physical "taint" inevitably attaches to a physical-thing statement when it is translated into sense-datum idiom. The gist of Blose's essay is that the physical-taint objection is essentially valid, but—. What follows the "but" in this exciting chase after an elusive truth makes for exciting reading and shows philosophical subtlety on the part of the author.—W. G.

FINDLAY, J. N. *Psyche and Cerebrum.* Milwaukee, Wisconsin: Marquette University Press, 1972. N.P.—This short, suggestive essay was the 1972 Aquinas Lecture at Marquette. It contains an outline of Findlay's critique of "mechanistic neuralism," i.e., belief in "invariant, isolable (neural) factors and rigorous laws governing their interaction." In a manner reminiscent of Bergson, he sees this view as the product of specifically intellectual activity which naturally produces a "world of remote objects, all fully interpreted, which stand over against our subjectivity. . . ." Speculative and experimental neurology thus present a view of mind in which the self is identified with the cerebrum. "The cerebrum . . . becomes the true man . . . to whom we attribute all our highest preceptive, cogitative, emotional and practical feats" (23). Findlay attacks no particular version of this view. His main concern is to argue against the tendency in thought. He holds that the difficulties faced by a through-going cerebralism are purely logical, and he sketches a number of arguments designed to expose some of the problems involved in reductionist accounts of perception, thought, and action. What makes the essay particularly interesting is the way in which it incorporates scattered but important insights of Wittgenstein, Husserl, and Hegel (among others) which bear significantly on any treatment of the mind-brain problem. What is frustrating is the pervasive generality of Findlay's characterization of "cerebralism." No cerebralist is identified, and no recent defense of the position is discussed or referred to in detail. Perhaps this is both unavoidable and appropriate in a public lecture. But it may leave the reader wondering how Findlay's critical points apply (or fail to apply) to recent important discussions of the problem.—J. W.

GILLAN, G. (ed.). *The Horizons of the Flesh; Critical Perspectives on the Thought of Merleau-Ponty*. Carbondale: Southern Illinois University Press, 1973. 195 pp. $7.96—This collection of eight critical essays makes a significant contribution to the secondary literature on Merleau-Ponty. As stated in the preface, the intention of the book is "to bring to expression the levels and directions through which the thought of Merleau-Ponty moved from *The Structure of Behavior* to *The Visible and the Invisible*." The first essay, by Gillan, entitled, "In the Folds of the Flesh; Philosophy and Language," sets the context for the essays which follow. It centers around the two "foci" of Merleau-Ponty's "struggle with the meaning of being": "the language of philosophy and its self-discovery within the corporeal texture of language itself, the flesh of language" (pp. 1–2). Gillan's comprehensive exposé traces the development of Merleau-Ponty's thought in terms of these two foci. Don Ihde's "Singing the World; Language and Perception" suggests a certain "priority" for Merleau-Ponty of language over perception. Language is "not just one dimension of being," and the problem of perception, consequently, has become "enigmatic" (p. 74). Ihde also raises the question of whether Merleau-Ponty has overcome the nature-culture dichotomy.

The third essay, Alphonso Lingis' "Being in the Interrogative Mood," contrasts Hume's epistemological skepticism with Merleau-Ponty's "interrogative thought," for which *being itself* is characterized by distance, vacillation, negativity, radiation, and delay. In his essay, "Merleau-Ponty and the Primacy of Reflection," Raymond Herbenick considers Merleau-Ponty's philosophical stance in his early works as a "phenomenological positivism," where reflection has an epistemological or methodological primacy over the unreflected. In a novel fashion, Herbenick uses J. L. Austin's distinction between "demonstrating the semantics of a word" and "explaining the syntactics of a word" to elucidate Merleau-Ponty's two methods of "radical reflection"—"living the body," and adopting "a new way of looking at things" ("intentional analysis") (pp. 99–101). The last four essays, then, broadly speaking, deal with Merleau-Ponty's "interrogation" of three other philosophers. Bernard Flynn presents Merleau-Ponty's critique of Sartre's *Being and Nothingness*.

Joseph Bien's essay, which discusses Merleau-Ponty's conception of history, and Dick Howard's "Ambiguous Radicalism: Merleau-Ponty's Interrogation of Political Thought" take up Merleau-Ponty's questioning of Marx. (Howard broadens the context to Western Marxism in general.) Bien considers Merleau-Ponty's sympathetic interpretation of the young Marx as well as his attempt to appropriate the writings of Max Weber. Howard, stressing the significance and pertinence of Merleau-Ponty's political thinking, takes him to task for supposedly misinterpreting Marx as regards the question of the "party." Ronald Bruzina, finally, treats Merleau-Ponty and Husserl in terms of the "idea" of science. Bruzina argues that Merleau-Ponty, despite his critique of Husserl, would have to presuppose an ideal of science. Both for Merleau-Ponty and Husserl, furthermore, science would have difficulty justifying its own foundations. In this regard, certain remarks of Merleau-Ponty in *The Visible and the Invisible* are compared to passages in Husserl's *Crisis* which have a voluntaristic (one might say, a Nietzschean) overtone. Man, it seems, is reasonable because he wants to be so.—D. E.

HINTIKKA, K. J. J., J. M. E. MORAVCSIK, and P. SUPPES (eds.). *Approaches to Natural Language: Proceedings of the Stanford Workshop on Grammar and Semantics*. Dordrecht, Holland: D. Reidel, 1973. 526 pp. Dfl 55—The approaches in question here are exhibited in examinations of specific problems, rather than surveyed or generally summarized. Most of the volume should interest philosophers. Recent linguistic theory has been torn between the generative semanticists, who fuse syntax and (logical) semantics in maintaining that "the rules of grammar are identical to the rules relating surface forms to their corresponding logical forms" (p. 197), and the interpretive semanticists, who find syntactic deep structure a well-defined notion and who believe that the semantic interpretation of sentences derives from inputs from several levels of linguistic structure. J. Bresnan, in "Sentence Stress and Syntactic Transformations," gives a clear and elegant version of her defense of one aspect of the interpretivist position. She argues that aspects of the stress pattern of sentences can be easily and compactly explained only if lexical items are inserted at the level of syntactic deep structure *before* the application of syntactical transformations. W. C. Watt's "Late Lexicalizations" argues the generativist position that the lexical peculiarities of natural languages tend to be introduced at various stages in the application of syntactical transformations. Bresnan's paper is particularly helpful to philosophers who want to make sense of linguist's current arguments: her evidential appeals, reasoning, and terminology can be grasped by someone with little background in technical linguistics. The volume also includes three papers, two by Hamburger and Wexler and one by Peters and Ritchie, on the abstract theory of grammar, which has come some distance since Chomsky's contributions. These papers follow out various aspects of the realization that, when abstractly considered, transformational, and even somewhat less powerful rules, are too powerful (without restrictions) to allow nonarbitrary solutions to the problem of identifying the grammars of particular languages.

The semantics section of this book contains some comments by Hintikka on some "misunderstandings" of logical notions that he finds in recent linguistic work. Most of this is merited; but one might question his dismissal of the claim of some linguists that the two sorts of reference philosophers find in modal and belief-attitudinal contexts are to be found quite generally in natural language and not in such contexts alone. A major feature of the volume is Richard Montague's contribution of one of his fuller versions of a semantic-based grammar explicating the quantification aspects of English. Aside from commentary about Montague grammars, papers by Moravcsik on mass terms in English, and by B. H. Partee on the semantics of belief sentences, attract the most commentary. D. Kaplan ends the volume with a suggestive, provoking, and bubbling piece on referential paradoxes titled "Bob and Carol and Ted and Alice."
—J. L.

JANIK, A., and S. TOULMIN. *Wittgenstein's Vienna*. New York: Simon Schuster, 1973. 314 pp. $8.95—Ludwig Wittgenstein concludes his *Tractatus* with the injunction, "What we cannot speak about we must pass over in silence." As the concluding proposition of a tersely written, tightly organized work, the reader would expect it to have a strong bite. Yet the statement has been variously ignored, dismissed, and misunderstood, interpreted as the inspired words of a mystic or as the final banishing of metaphysics from philosophical discourse. It is with the help of Janik and Toulmin's work that it becomes clear how the proposition serves as the crown to a book which Wittgenstein maintained was primarily an ethical work. In presenting the Viennese Weltanshauung of the late nineteenth and early twentieth century, there emerges the picture of a society where appearances ruled in all areas of cultural life, in the government, and in the arts. Viennese society was characterized by a vast impotent bureaucracy, the Strauss waltz, and the feutillion. The leader in the inevitable reaction was Karl Kraus. He instigated a critique of Viennese society through an ingenious and refined use of cultural modes of expression. Along with him, leaders emerged for each of the special arts, language, music, architecture, painting, sculpture, who, in their own particular role, tried to restore truth and responsibility to the affairs of men. The development of the characters and the issues involved make this book important. Of lesser value, unfortunately, are the chapters dealing specifically with the work of Wittgenstein, who is represented as one of the emergent leaders. Although we are rhetorically persuaded that his talk of "simples," "representation," "depicting," or "language games" is tied up with the critique of a fundamentally sophistical intellectualism, we are not led on to seeing how the critique was carried out.—W. A. F.

JASENAS, M. *A History of the Bibliography of Philosophy.* Hildesheim: Georg Olms Verlag, 1973. 188 pp. N.P.—This book is designed to demonstrate that "the bibliography of philosophy has not emerged directly from a barbaric past; it has a long history . . ." (p. 137). It begins with the first known printed bibliography, that of Frisius in 1592, and works its way methodically to 1960. By sketching the contents and divisions of these bibliographies, Jasenas provides us with evidence of what philosophers of different eras took philosophy to be. Some bibliographers were professional philosophers and some were not. But it is clear that those who were not philosophers generally consulted philosophers in preparing their bibliographies. Particularly interesting is the way in which Jasenas ties a given bibliography to the cultural, intellectual, social, political and even musical happenings of the day. This is especially the case when he offers an explanation for the absence of one or other book in a given bibliography. He is careful to note the religious and philosophical (e.g., anti-Aristotelian, pro-Aristotelian, etc.) preferences of the bibliographer. Granted there is a good deal of speculation and likely guesswork on Jasenas' part, his suggested explanations give to his book an importance which it would have lacked without them.

Jasenas points out that bibliographical work was almost exclusively a German phenomenon until the twentieth century and Protestant until the nineteenth. In most cases he follows the format of life-content/arrangement-evaluation. These evaluations provide quite interesting accounts of the genesis of the newer components of contemporary philosophy: philosophy of language, philosophy of history. The turbulent career of metaphysics is very well covered. The book contains a bibliography of non-bibliographical works together with a bibliography of bibliographical works and a "short-title list of major philosophical works discussed in standard histories of philosophy." Though by no means an exhaustive list, the latter is particularly helpful in that it gives complete and accurate bibliographical data on first editions of works from the fifteenth to the twentieth century. The book has at least a threefold value for the historian of philosophy: first, it links bibliographical studies with the intellectual milieu of the period; second, it makes manifest what philosophy progressively came to mean and how it was to be divided; third, it is a very helpful research tool for gathering information on a given philosopher.—D. A. C.

KAHN, C. H. *The Verb "Be" in Ancient Greek.* "The Verb 'Be' And Its Synonyms: Philosophical and Grammatical Studies," Part 6. Dordrecht, Holland: D. Reidel Publishing Company, 1973. 486 pp. $39.00 —The goal Kahn sets for himself in this impressive and important book is "to give an account of the ordinary, nontechnical uses of the Greek verb [*eimi*] . . . by dealing extensively with the earliest evidence (from Homer) and by referring to parallel evidence in cognate languages" so as to "make this a study of the Indo-European verb *be*" (pp. ix–x). He uses a modified version of Zellig Harris' transformational grammar (explained in Ch. I and applied to Greek in Ch. III) for analyzing the copula, existential and veridical uses of the verb *be* in Chs. IV, VI, and VII, which are the core of the book and which will be of special interest to students of linguistics. Ch. II on the concepts of subject and predicate, Ch. V on the general theory of *be* as copula, and Ch. VIII on the conceptual unity between the multiple uses of *be* are geared "for readers whose interest is primarily philosophical."

Neither linguists nor philosophers will be disappointed, as is clear from samples of Kahn's thought-provoking statements. "The definition of kernel sentences [the sentence-forms of the language, from which all other sentences may be derived by grammatical transformation] represents the most important contribution to the philosophical search for a ground-level 'object language' *within* natural languages since Aristotle's account of the basic forms of predication in his *Categories*" (p. 17). "It has often been supposed that a substance-attribute metaphysics is a projection onto the world of the noun-verb or subject-predicate structure of sentences in Greek and cognate languages." No, "the appearance in many or most languages of a noun-verb distinction, and hence of a subject-predicate sentence structure as well, is the reflection within grammar of certain fundamental conditions underlying all human use of language" (p. 49). In its existential uses *eimi* has four lexical nuances: vital (to be alive), locative (to be present here), durative (to last, endure), properly existential (e.g., "There are some who laugh"), each of which can receive further meanings in six different syntactical constructions. The last of these ("There are gods but there is no Zeus") is philosophically significant but is post-Homeric; its originality is that the verb *esti* poses the extra-linguistic subject as such, whereas in the other types it poses "an extra-linguistic subject of a given sort that satisfies" the conditions formulated in subsequent phrases or clauses (e.g., "There is a certain Socrates, who is wise. . . ."; p. 301). But neither this nor any of its five other existential meanings of *eimi* is primary and original. According to Kahn's "modest Copernican Revolution" its use as copula is primary; its existential function is secondary and derived (p. 394). Indeed, "every existential use of *eimi* is second-order *precisely to the extent that it is existential.*"

The copula construction of *eimi*, together with its existential and veridical uses, issued into a single concept of Being, which is not univocal but a *pros hen* equivocal: "the verb has a number of distinct uses or meanings that are all systematically related to one fundamental use" as copula or sign of predication (p. 401). But the Greek doctrine of Being from Parmenides onwards is not that of Heidegger or any other existentialist because Greek ontology focussed on the third-person and non-personal form, these latter on the first-person form. In fact, "the impersonal and objective construal of ontology" of the early Wittgenstein and of Quine with their preference for states of affairs and objects described in third-person language stands "closer to Greek interests than do the continental philosophers of Being" (pp. 416–17). Such a sampling of insights indicates that no linguist or philosopher can afford, despite its high price, to be without Kahn's splendid book.—L. S.

KAULBACH, F. *Einführung in die Metaphysik.* Darmstadt: Wissenschaftliche Buchgesellschaft, 1972. 244 pp. N.P.—The fulfillment of the pedagogical intention of this work takes the form of developing a specific conception of metaphysics. The author's conception is not the original one in terms of the object of metaphysics, but one that encompasses as well the various philosophical developments that affect the original conception of it, including critiques of it. The author conceives metaphysics in terms of "motives" of thought, originally articulated by Plato, namely, 1) the knowledge of principles of being; 2) the standard of true knowledge; and 3) the way in which this knowledge is attained. According to the author, the first definitive treatment of the first of these themes is given by Aristotle, in the conception of the essences of things as the principles of being (Chapter 2). In subsequent treatments, these themes undergo various transformations, which are presented by the author as being the result of critical reflections on prior treatments. In fact, they are the result of shifts in perspective, the most fundamental of which is, as the author himself notes, the introduction of perspectivity as a category in the sense of a condition of knowledge. The author traces the function of this category in the thought of various philosophers from Augustine to Nietzsche, suggesting, though not explicitly establishing, its theological origin (Chapter 3).

He then formulates the current philosophical task as the reconciliation of the Aristotelian conception of the principles of being with the modern scientific conception of them as laws, a conception of them which is an outgrowth of the perspectival attitude. The theoretical aspect of this task is the formulation of the tension between the two conceptions of being in terms of their origin in the two antithetical conceptions of nature as free, i.e., uncontrolled by human agency, and as subject to mastery by man (Chapter 4). For the author, the tension is resolved not on the theoretical but on the practical level, in so far as the technical imposition of the laws of nature propounded by man on things creates their essence (Chapter 5). The author then turns to questions of method, the ultimate one of which is the question of the possibility of a comprehensive system (read: perspective); on the basis of an analysis of nineteenth century systems, the author answers this question in the negative (Chapter 6). Finally, (Chapter 7), the author analyses three critiques of metaphysics, the Kantian, the positivistic, and the Diltheyan, showing that they, too, exhibit the "motives" in terms of which the author conceives metaphysics.—W. A. S.

KOLENDA, K. (ed.). *A Symposium on Gilbert Ryle, Studies In Philosophy*. Houston: Rice University Press, 1972. N.P.—An outgrowth of Ryle's three week visit at Rice in the spring of 1972, this collection of critical essays bears some resemblance to the collection edited by Oscar P. Wood and George Pitcher in the Anchor series. The principle differences are: 1) the range of topics treated here and the detail of treatment is considerably less extensive than in the Wood collection, and 2) this volume contains two new essays by Ryle himself: "Thinking and Self-Teaching" and "Thinking and Saying." Four papers by members of the philosophy staff at Rice form a group. Each of them discusses Ryle's contribution to a problem in which the author is interested, carefully delineating Ryle's analysis and treatment of the problem. Generally Ryle is regarded as having said some important things about the topic, having said some things which are questionable, and (in spite of the questions) having at least advanced the topic in an important way. Subjects dealt with in roughly this fashion are "Reference and Existence" by Lyle Angene, "Sensations, Feelings, and Expression" by Richard J. Sclafani, "Dispositions and Hypotheticals" by Robert W. Burch, and "Why Virtue Cannot be Taught" by Thomas McElvain. As surveys of Ryle's position and as assessments of his views, the essays are consistently good.

The papers by Kolenda and Bouwsma present over-views of Ryle's thought. In a postscript Ryle takes some exception to each. He mildly resists Kolenda's suggestion that in *The Concept of Mind* he "was engaged in a task of heroic knight-errantry on behalf of the oppressed concept of Man." His objection to Bouwsma is sharper and particularly interesting because it provides a rare glimpse of Ryle's view of Wittgenstein—a view clearly at odds with the interpretation of Bouwsma, whose essay compares Wittgenstein and Ryle. The essay by Aldrich and the essays by Ryle himself fall into a different category. In each case a philosopher is developing arguments, themes, and topics developed extensively elsewhere. For those who have followed Aldrich's thought on the problem of perception, this essay will be particularly interesting. He aims to put the concept of intentional object in a new light, and in the process distinguishes his position from Dretske's and Anscombe's, among others. Ryle's contributions are part of a series of essays on thinking. A basic contention is that the view that there is "no thinking without vehicles" can be refuted. At least through the 1968 "The Thinking of Thoughts," Ryle seemed unable to provide the supposed refutation. In these essays he addresses himself to this question in much greater detail and with more, if not final, success.—J. W.

KRIKORIAN, Y. H. *Recent Perspectives in American Philosophy*. The Hague: Martinus Nijhoff, 1973. 90 pp. G 17.50—Is there a distinctive American philosophy of the twentieth century? If so, what are its defining features or constitutive attributes? Krikorian, who himself has helped to shape and steer American philosophy, answers Yes to the first question and gives a clear and revealing answer to the second. He declares bluntly that "whereas continental philosophy is existential and British philosophy is analytical, American philosophy is empirical." Acknowledging, then, that the empirical temper "is not something that belongs exclusively to America," he pinpoints the brand of empiricism that is dominant in the United States. Empiricism in America, as Krikorian sees it, is the view which favors "objectively and socially verifiable pronouncements," or "confirmation through demonstrable evidence." It applies this criterion in value experience as well as in science, business, and technology. It contains within it threads of pragmatism, experimentalism, realism, naturalism, functionalism, temporalism, logical positivism, and phenomenology.

American empiricism, according to Krikorian, "is not opposed to reason," despite the traditional rivalry between empiricism and rationalism. Indeed, the prominence of reason, logic, and clarity in the empiricist endeavors and goals of American philosophers is reflected in the titles of two of Krikorian's seven chapters devoted (after a general chapter) to individual thinkers. Those two chapters are on "Cohen's Rationalistic Naturalism" and "Blanshard's Rationalistic Idealism." Idealism? How did this come in if realism is a dominant thread in American philosophy? It came in because Blanshard's world view is not the subjective idealism which is commonly taken as realism's contrary, but is rather an attempt, thoroughly consistent with empiricism, "to interpret reality in terms of human aspirations and . . . higher human categories" and, from another angle, in terms of "a network of logical relations . . . forming a single system or Whole." The philosophers who receive one chapter each in Krikorian's illuminating treatise are Dewey, Cohen, Singer, Hocking, Blanshard, Whitehead, and Sheldon. Each of these chapters except the one on Blanshard was previously published. A useful subject index is included.—W. G.

KRISTELLER, P. O. *Renaissance Concepts of Man and Other Essays.* New York: Harper & Row, 1972. 183 pp. N.P.—The following papers are contained in this book: "Renaissance Concepts of Man: 1) The Dignity of Man; 2) The Immortality of the Soul; 3) The Unity of Truth" (these three papers being the 1965 Arensberg Lectures); "Italian Humanism and Byzantium;" "Byzantine and Western Platonism in the Fifteenth Century;" "Renaissance Philosophy and the Medieval Tradition" (the 1961 Wimmer Lecture) and, finally, "History of Philosophy and History of Ideas." All of the essays have been made public, although, to my knowledge, only the last four papers ever appeared in print. The fourth and fifth papers have been published (in Italian) but without footnotes. In his Preface, Professor Kristeller indicates that "the second three essays, in describing the Byzantine and medieval backgrounds of Renaissance thought, offer a supplement to my first Torchbook, *Renaissance Thought* (1961) in which the classical sources of Renaissance thought are discussed." The last paper appeared in the *Journal of the History of Philosophy* (1964).

My remarks will concentrate on the first three papers which have not appeared in print, most particularly on the second paper, dealing with the immortality of the soul. Kristeller confesses at one point that "I am partial to it [Platonism]; and that the manner in which I understand and describe it, and even the historical importance I attach to it may be influenced by this partiality" (p. 148). And although Kristeller does speak up for a particular brand of Aristotelianism (Paduan Averroism), it is not possible to say from this that his views are Aristotelian. We should take him at his word. With regard to the importance of the question for the sixteenth century of the immortality of the soul, Kristeller would have it that Platonism, almost singlehandedly, was responsible for the upsurge of interest in it. Negatively, he argues that immortality was not a very important item for Aristotle and for his pagan, Moslem, and Christian followers. By way of example, Kristeller cites Thomas Aquinas who "duly defends the incorruptibility and future beatitude of the rational soul, but he seems to avoid the term 'immortality,' and he does not attach especial importance to the subject" (p. 29). In the first essay, Kristeller attempts to show that the theme of the dignity of man found in his *Oration* is not simply a rhetorical exercise for Pico, but a very deeply cherished belief. Kristeller argues that in the *Heptaplus* the very same points are made regarding the dignity of man. But where one would expect footnote citations referring to the *Heptaplus*, Kristeller for some reason cites *De Hominis Dignitate*, which defeats the whole purpose. Aside from some very minor objections, Professor Kristeller's collection of Renaissance papers is a very fine book and should be quite helpful in history of philosophy courses which require a brief but careful survey of the Renaissance period.—D. A. C.

L'Archevêque, P. *Teilhard de Chardin: Nouvel Index Analytique.* Québec: Les Presses de l'Université Laval, 1972. 289 pp. $5.00— In 1967 Paul L'Archevêque published an analytical Index of the works of Teilhard de Chardin which has become an invaluable reference source for the thought of the renowned French scientist. This new Index is to a great extent a continuation and implementation of the former one. While it makes up for some of its deficiencies, it contains many additional references to Teilhard's works which appeared after 1967, including some unpublished letters and pertinent material from the "Cahiers" of the "Editions du Seuil." A simple look at some of the entries, such as "Christ" with 119 references, "Homme" with 277, "Matière" with 145, and "Vie" with 198, gives an idea of the throughness and practical value of this new Index. Thus one can only wholeheartedly agree with Claude Cuénot, the well-known Teilhard scholar and biographer, who, in addition to admitting that both he and the "Fondation Teilhard de Chardin" have greatly profited from the 1967 Index, recommends very highly the *Nouvel Index Analytique* and expresses the wish that at some future date the two Indexes be combined into one volume with an additional Index of Names (*Preface*). Cuénot's suggestion has been favorably accepted by the author, who intends to carry out the project as soon as all Teilhard's works are published (*Avant-Propos*). —B. M. B.

Lynch, J. P. *Aristotle's School.* Berkeley: University of California Press, 1972. 247 pp. $10.00—Werner Jaeger's epic-making work on Aristotle long ago established that the form and substance of the various types of Aristotelian *logoi*, or treatises, are historically unique in that their intelligibility is indissolubly connected with the Lyceum as an educational institution. The laborious reconstructive work of centuries of commentators should not obscure the fact that both the exoteric and the esoteric treatises have their ultimate *Sitz im Leben* in the Lyceum, that peculiar philosophical school whose communal life formed, perhaps, the first historical semblance of a "university" in the modern sense. Professor Lynch studies Aristotle's school from its pre-Aristotelian origins as a religious sanctuary in Athens to the expiration of the Athenian Peripatos at the beginning of the first century before Christ. Although his primary focus is not on the relation between the nature of the school and the kind of philosophical thought it produced, his scholarly book reinforces the conviction of the historian of philosophy that investigating the Peripatos as a community of men concerned with higher education is a necessary propaedeutic for understanding how the *Perpatetikoi* actually philosophized. In a famous essay on the Academy and the Peripatos Hermann Usener discerned that the essence of these philosophical schools resided in their *geistige Arbeit*.

Professor Lynch interprets the philosophic act too narrowly and underestimates the value of his book for philosophers when he states that his research will have little utility in "understanding Greek philosophy as a speculative phenomenon" since his principal concern is with "what Aristotle and other Greek philosophers did as teachers. ..." He claims that *philosophia* as developed in the Athenian schools of the 4th century B.C. possessed a specific dimension that can be described as "higher education," which he equates with praxis exclusively. Failure to clearly distinguish theory from practice is conceivably a methodological error for the historian of education, but the coalescence of just these factors is indispensible to any adequate historical or essential definition of philosophy. By concentrating on a significant and historically complex phenomenon Professor Lynch's fine piece of classical scholarship enables the student of philosophy to see that philosophy is indeed a product and reflection of the totality of all objective interacting conditions of life.—C. F. B.

MORICK, H. (ed.). *Challenges to Empiricism.* Belmont, California: Wadsworth Publishing Company, 1972. 329 pp. N.P.—The fifteen selections in this volume are collected around the thesis that many of the foundations and tenets of empiricism are mistaken and must be either rejected outright or radically revised. To introduce these essays, Morick briefly traces the development of modern empiricism from what he considers its source in Hume's theory of knowledge through the phenomenalist stage to the present conception of empiricism, one of whose basic principles continues to be the fundamental role of observation in the acquisition of knowledge: perception provides the foundation of empirical knowledge. It is particularly this "foundation picture" which has been denounced by many contemporary critics of empiricism as both oversimplified and misleading. The critics, including those in philosophy of language, philosophy of science, and scientists themselves, have proposed three related objections to this notion. The first is that rather than drawing a sharp distinction between observation claims and theoretical claims, observation claims must be understood in terms of a network of background assumptions. The second objection holds that it is this network of assumptions which bestows the very meaning on observation claims. Consequently, an observation term such as "red" will not maintain a fixed meaning throughout various changes in theories. Further, since the objection also covers observation words, the notion of ostensive words is rejected. The final objection is that our observations themselves should be looked upon as interpretations arising from these background assumptions rather than merely barren perceptions, isolated from one's beliefs and preconceptions. The selections, most of which are well-known, specify these and other critical arguments and revisions. They are arranged into three slightly overlapping sections: "Empiricism and Ontology" (Carnap, Quine, W. Sellers, and Putnam), "Empiricism and Science" (Popper, Feyerabend, Kuhn, and Hesse), and "Empiricism and Linguistics" (Chomsky, Edgley, Goodman, and Fodor). Considering the purpose and scope of the book, the contributions have been appropriately chosen. An eight-page annotated bibliography is included.—B. G. H.

PRICE, H. H. *Essays in the Philosophy of Religion.* Oxford: At the Clarendon Press, 1972. 125 pp. $7.75—This book is based on the Sarum Lectures given at Oxford in 1971. Though the text is revised, it retains much of the character and tone of an oral presentation. The general theme concerns interactions between studies of religious experience and of paranormal phenomena. As Price says, "I have tried to describe how a philosopher who is interested in psychical research might approach some of the problems of religion, and to consider what insight we can derive from this approach" (p. v). His topics include the combination of love and fear in religious experience; paranormal cognition, symbolism, and inspiration; views of petitionary prayer as well-wishing and as wishful thinking, together with a possible reinterpretation based on self-suggestion and telepathy; the notion of latent spiritual capacities and their affinities with paranormal capacities (e.g., the inward search for the divine); a survey of motives which especially favor disbelief in life after death (e.g., quality of life vs mere duration); and conceptions of the next world in terms of embodied or disembodied survival, which suggest a convergent third or mediumistic version characterized by "embodiment" in the form of mental images and some consequent intersubjectivity. A brief appendix offers some comments on the post-resurrection appearances of Jesus.

No overall conclusion is drawn. Price seems content to mark out affinities between his two areas of interest and to suggest insights which might be contributed from one to the other. For example, he contends that barriers between insiders and outsiders in both areas might be overcome by distinguishing between "believing a hypothesis and acting (inwardly) as if it were true" (p. 74). However, the dominant flow of insights is from psychic research to philosophy of religion. For philosophers, there is an interesting sketch of the development from synthetic to analytic of the *concept* of the Deity, though it is marred by the claim that the results are, or must be, "the" theist's position.—C. A. C.

REEVES, M., and B. HIRSCH-REICH. *The Figurae of Joachim of Fiore.* Oxford: Oxford University Press, 1972. 350 pp. plus 12 pages of plates. $32.00—Although Joachim of Fiore (1135–1202) created a rather intriguing theology of history along with a primitive theory of hermeneutics (which is curiously blended with his theology of history), his importance for the historian of philosophy is most likely to be in his reaction to the trinitarian doctrine of Peter Lombard and in his influence on Bonaventura. Joachim invokes what he calls a *spiritualis intellectus* against the teaching of Peter Lombard. This spiritualis intellectus "includes both the preparation of arduous study and the experience of mystical illumination. It includes, thirdly, the phase of close intellectual work which follows the illumination, in which the 'given' clues are used to organize the gathered material into patterns which now emerge from within" (p. 3). This *spiritualis intellectus,* obviously mystical, is quite eschatological and numerological. Two, three, five, seven and twelve figure prominently and often in Joachim's account of the panoply of history. The specific account of Joachim's reasons for his opposition to Peter Lombard is quite interesting and, historically, very important (pp. 219–20). In this passage Joachim is linked with the anti-dialectical tradition which, led by St. Bernard of Clervaux, found scholastic theology to be static, dry and alien to spiritual experience. Scholastic theologians are following nothing but the *carnalis intellectus*.

Although Bonaventura had some unpleasant encounters with Joachimites, it still is possible to say that "some parts of his writings show him to have been thinking in categories very close to Joachim's" (p. 300). Bonaventura's use of the tree metaphor, his theology of history, his use of generation and fructification and his notion of the *requies animarum* can to some degree be linked to Joachim. But on at least two points, i.e., trinitarian doctrine and the proximity of this *requies animarum,* Bonaventura is careful to avoid sounding like Joachim (pp. 301–04). Finally, there is a chapter dealing with Joachim's influence on Dante's symbolism. To a great extent the chapter is concerned with a critical review of Tondelli's conclusions on this point, but there are many new suggestions in the chapter. —D. A. C.

ROTH, R. P. *Story and Reality.* Grand Rapids, Michigan: William B. Erdmans, 1973. 197 pp. $3.45—There are fresh currents running through this volume, subtitled "An Essay on Truth," which dispel some accumulated but unexamined theories: e.g., that St. Paul took literally the three-story picture of the world; that nature can be subsumed under the category of history so that all meaning is historical; that there is a genuine dilemma between absolutism and relativism in morality. The author argues that the clue to reality is "story" for the simple reason that reality itself *is* story: a dramatic conflict between persons, ambiguous at its core. The real then is grasped as story, rather than in terms of philosophical *Weltanschauung*, or scientific *Weltbild*, or history which in furnishing elements to the story distinguishes it from myth. Beyond emotive propositions, verbal propositions, and descriptive propositions lies a fourth kind of statement proper to story. Unfortunate oversimplifications compromise the book's thesis: e.g. that Logos, Substance, and Life-Process are concepts whose use is metaphorical as is that of causality; that Aquinas taught grace is a supernatural substance; that Aristotle viewed reality as ideal rather than actual; that to conceive of evil as non-being is to deny it reality, to view it as illusory. At work here is a decided antipathy to the narrowings of rational discourse that rules out anything that might undergird the author's "story" and give it ontological density.

However, if story is the clue to reality, then the search for truth is most fruitfully pursued in literature: the tale that is told reveals the tale that is so. Here the volume offers its richest yield in Roth's attempt to concretely illustrate his thesis. From within the framework of Western literature he explores five themes: evil, love, holiness, hope, and meaning in ways that are never forced, always richly imaginative and illuminating, and finally convincing. Still the problem remains: how does meaning come to be given to the elements of the story, and how is any interpretation authenticated? Not by history for "historical details get mixed up in the story, but the latter is the final authority" (p. 30). Roth's answer is simply "faith," in the sense of credulity in a story by which a culture lives, ultimately faith in the basic story underlying and informing all versions of the human story, that comes to its fullest articulation in the Christian Story, viz., the New Testament. But such faith is merely "given" and Roth eschews any critical grounding of it. The objectivity of the data of faith is inferred from the experience which is self-authenticating. In the end story is left to justify itself.—W. J. H.

RUSSELL, D. A. *Plutarch.* New York: Charles Scribner's Sons, 1973. 183 pp. $7.95—This is the best general book available in English on Plutarch, written by the foremost scholar at Oxford in the Greek literature of this period, containing many fresh insights, along with a clear grasp of the subject and accurate scholarship. The scope is wide: Plutarch's life; the language; style; and form of his writings; his sources and method of writing; his philosophy and religion; the moral essays and his humanism; the nature of the *Lives*; and the inevitable *Nachleben*. The size, unfortunately, is small.

Russell's best chapters are on Plutarch's style, with its baroque fantasies sometimes inspired by Plato's vision of the world beyond, and the chapters recreating the literary world and cultural values of the time. Through Russell we can come to an appreciation of the era from the end of the Second to the beginning of the Third Centuries A.D. when a romantic revival of the past attracted by the pure Attic of Plato's literary classics stimulated an interest in Platonic philosophy. His treatment of Plutarch's more metaphysical works, however, leaves something to be desired. He correctly sees Plutarch as a somewhat unorthodox Middle-Platonist, but he does not push him too far into Neoplatonism. His comments on Plutarch's interpretation of the world soul do not give a clear enough picture of Plutarch's method of interpreting Plato. In this case the use of isolated passages from dialogues other than the *Timaeus* led to an ingenious but illegitimate and anthropomorphic description of the world soul. Russell tends to overlook Plutarch's conception of the soul as mindless rather than evil. Further, he is not strong in noting the deviationism in so far as Plutarch confused or identified God, the Forms, and the soul, as well as God, the Good, and the One, and spoke of the phenomenal world as a mirror of the Ideal, while being the first to suggest the hypostases of being which became prominent in the Gnostic writings. His treatment of Plutarch's demonology (a crucial issue in Neoplatonism) is sane.

One can happily cite him on Plutarch's religious writings: "The Platonist theme of the essential reliability and goodness of God is the key to Plutarch's attitude toward religion."—belief without dogmatism (p. 81). Much can be learned from his comments on the moral essays: the exercise of virtue as the key to moral improvement; stress on the gentler forms of *arete* as the essence of the good life; stress on the education of women to improve society; and that *philanthropia* and stress on the mean as the heart of his ethical teaching. Finally one can cite his observation on the *Moralia*: "He lavished art and ingenuity on these elaborate set-pieces, in which imitation of Plato did not prevent him from adumbrating doctrines which Plato never knew and creating fantasies in the taste of his own age" (p. 75).—F. E. B.

SMITH, J. E. *The Analogy of Experience*. New York: Harper and Row, 1973. 140 pp. $6.95—The author was moved to write this book by reason of his deep concern over contemporary man's loss of a sense of God's reality. This situation can be partly attributed to the way in which faith has been presented. Proclamation merely confronts the mind of the hearer; it does not penetrate the total life-pattern and experience as religious truth demands. Interpretation is necessary so that the listener may see the truth in its concreteness. Professor Smith, in the tradition of Pierce, James, and Dewey, holds for a reconstructed view of experience which is the cumulative and meaningful result of a many faceted encounter between a concrete person and all of reality. Experience has many dimensions, moral, aesthetic, scientific, and religious, which give the life of a person direction and purpose. Some of these experiences are of such a kind as to carry with them direction and purpose in regard to life as a whole. He argues that religion is at the center of experience; and experience is at the center of religion. Professor Smith sees himself in the tradition of "faith seeking understanding." He does not attempt to provide a proof for God's existence because religion does not really *live* at the level of proof. Religion, however, cannot remain living without understanding and commitment to those ideas which are morally fruitful and illuminating of experience. Christian conceptions must be presented in experiential terms. Otherwise they will neither be understood nor accepted. This approach requires that we begin with man rather than, as has been traditional, with God.

Because of his finitude, similarity and analogy are necessarily employed by man, in his effort to speak about God. While previous discussion of analogy has contributed much to the understanding of religious language, the attempt remains abstract precisely because of its emphasis on language; rather than on experience, which language is supposed to represent. The "analogy of experience" is put forth by Smith as a more apt medium for conveying religious insight since experience is at the heart both of life and religion. Justification for an "analogy of experience" is, however, to be found in an ontological foundation from which experience, in its nature and function, flows. Selfhood is the only available concept that might suggest what an infinite transcendent being might be like. Within our experience, the self is the only reality that we can think of as being related to everything and consequently, as being in some real sense an appropriate model of what an infinite God. The analogy of the self as experienced is similarly presented as a model appropriate for clarifying the human situation. This is designated as the "circular predicament" insofar as man is led to look for liberation to Christ, who is the experiential model of true selfhood. Professor Smith's work is a step in the direction of a broad and balanced philosophy of religion. It allows for the possibility of a transcendent being and for the possibility of a revelation and of the acceptance of it by man.—A. G.

TAYLOR, R. *With Heart and Mind*. New York: St. Martin's Press, 1973. 147 pp. $6.95—Taylor believes that if we "penetrate the illusions that encompass us," then we can see the picture of man as one with "God the creator." This picture is created by Taylor through his critical and sometimes whimsical approach to man's relations with himself, others, the world and God. What man must realize is the openness of creation. He must avoid the problem of intellectualizing or showing no feeling "for the sticks, stones and grass at [his] feet." Taylor feels that man is more concerned about his deeds rather than his being. Man has to rid himself of the "separateness of illusion, of failing to see what is in fact before us." The purpose of life is "not to do but simply to be." Within God's world there is death and deterioration yet there is an ever renewing generation of life in all its forms. This book offers a vision of creation that all can share.—G. D.

WEINGARTNER, R. H. *The Unity of the Platonic Dialogue.* Indianapolis: Bobbs-Merrill, 1973. 205 pp. $7.50—With this book Professor Weingartner has added to that portion of Plato-interpretation which attempts to illuminate the fact that "Plato wrote dialogues." His central claims are two: 1) Plato's argumentation cannot be understood outside the dialogue form within which Plato himself never appears; and 2) the unity which suffuses each of the dialogues can render potent the argumentation which would be otherwise either inaccurate or inadequate. Correlative to these theses, he argues, perhaps too briefly, against those who would try to ignore the dialogue genre altogether. For such interpreters, Weingartner's argument goes, ". . . questions and objections that are interjected by the other participants in the conversation . . . are seen as mere literary devices which provide Plato's prolocutor with plausible opportunities for stating and clarifying his own views." On the other hand, he criticizes those who would make of Plato nothing but a literary figure, and of Socrates a "god of refutation." Fittingly, though, Professor Weingartner also clarifies the more subtle mistake of reading the dialogues ". . . as *tranches de vie philosophique,* in which the *character* of Socrates plays the leading role . . ." (italics his). Having stated what he takes to be improper interpretive devices, he selects the *Cratylus, Protagoras,* and *Parmenides* and ". . . attempts to show, in some detail, that each of them can, and therefore should, be understood as a unified whole."

Historically, these three dialogues have proven refractory to satisfactory interpretation along the lines suggested by Weingartner. His claim is that an underlying theme can be found in each of these three dialogues which will unify otherwise incommensurate parts. In the *Cratylus,* "Hermogenes and Cratylus maintain theories of naming which, were they sound, would make dialectic impossible. Plato's aim is to keep the way open to dialectical inquiry." In the *Protagoras,* "Plato pits two conceptions of morality and education against each other and shows, by means of a dramatic interchange between Socrates and Protagoras, that the very problems the Sophist aims to solve require the philosophical methods and commitments of Plato's Socrates." And in the *Parmenides,* Plato "is engaged in revising his views on *both* the method by which knowledge is attained and on the nature of the objects of knowledge themselves." While the program of Weingartner's book is praiseworthy, the degree to which he carries out his own interpretive canon is oddly lax. Perhaps the lapses are unavoidable because of his opinion that "The drama of a Platonic dialogue can be concentrated entirely in its arguments, as it is in the first part of the *Parmenides*." But no real effort is made, for instance, to illustrate the unity of the hypothesis-section of that dialogue with the arguments of the first part, either according to an underlying theme or otherwise. In short, then, while this book will well repay study, it has fallen into the trap of interpreting Plato according to one's previously held opinions about what is and what is not important in the Platonic dialogues. When this is done, the unity of a Platonic dialogue becomes inaccessible.—J. R.

CURRENT PERIODICAL ARTICLES

PHILOSOPHICAL ABSTRACTS*

AMERICAN PHILOSOPHICAL QUARTERLY
Vol. XI, No. 1, January 1974

Moral Responsibility, Freedom, and Compulsion, ROBERT AUDI

    This paper sets out and defends an account of free action and explores the relation between free action and moral responsibility. Free action is analyzed as a certain kind of uncompelled action. The notion of compulsion is explicated in detail, and several forms of compulsion are distinguished and compared. It is argued that contrary to what is usually supposed, a person may be morally responsible for doing something even if he did *not* do it freely. On the basis of the account of free action, it is also argued that freedom and determinism are compatible and that, though a person is morally responsible for doing something only if he could have done otherwise, determinism does not entail that no one ever can, in the relevant sense, do otherwise. The concluding part of the paper suggests that, if the account of the relation between free action and moral responsibility is correct, then the class of actions for which we bear moral responsibility is significantly wider than a great many people suppose.

The Aboutness of Emotions, ROBERT GORDON

    I attempt to show that when someone is, e.g., angry about something, the events or states that conjointly are causing him to be angry conform to a certain structure, and that from the causal structure underlying his anger it is possible to "read out" what he is angry about. In this respect, and even in some of the details of the structure, my analysis of being angry about something resembles the belief-want analysis of intentional action. The chief elements of the causal structure I describe are a belief and an attitude so related in content as to constitute either a wish-frustration (in the case of negative emotions) or a wish-satisfaction (in the case of positive emotions). The analysis makes otiose, in those cases for which it is a correct analysis, the mysterious non-causal relation between an emotion and its "object" which is invoked by the majority of philosophers now writing on emotions.

---

    *Abstracts of articles from leading philosophical journals are published as a regular feature of *The Review*. We wish to thank the editors of the journals represented for their cooperation, and the authors of the articles for their willingness to submit abstracts. Where abstracts have not been submitted, the title and author of the article are listed.

CURRENT PERIODICAL ARTICLES

Epistemic Defeasibility, MARSHALL SWAIN

Knowing and Meaning, BARBARA HUMPHRIES

Evaluative Asymmetry, A. F. MACKAY

Making Sense of "Necessary Existence," B. R. MILLER

It is often claimed that "necessary existence" makes sense because, if "God exists" were true, it would necessarily render senseless any further question of the form "Why is it that such-and-such exists?" In this article I sketch how such a view arose, and criticize it for selecting a proposition that cannot do even the limited job allotted it. The basic reason is that no predicate (not even "____exists") is necessarily predicated of anything. Subsequently I argue that the job could be done by a proposition in which "exists" does not function as a predicate. This is the logically simple proposition "Exists," which is not only the conclusion of the Contingency Argument but has the additional merit of showing how necessity could be ascribed to existence quite properly, and not merely in some oblique fashion.

The Temporal Dimension of Perceptual Experience: A Non-Traditional Empiricism, DAVID M. JOHNSON

I think of data of perception not as objects of immediate awareness, but as givens *for knowledge*. Formal elements in what we perceive (e.g., visual shape) are more informative about the intersubjective world than are content elements (e.g., colors). For this reason, I identify the former with the important empirical part of perceptual experience, or "sensuous data." One controversial implication of this idea is that since formal elements in experience can extend over time, it must also be possible for sensuous data to extend over time. But how can one perceive (as opposed to remember) something which is not confined to a specious present? I argue that memory in addition to the senses must sometimes "take" that which is given from the external environment, because sometimes the only thing sufficient to distinguish the states of affairs which we learn about non-inferentially by means of perception is a whole temporal pattern of experience.

The Analogical Argument for Knowledge of Other Minds Reconsidered, THOMAS OLSHEWSKY

The analogical argument is not appropriately construed as an argument, since the role of analogy in inquiry is in discovery procedures and not in justification procedures. Nor is it even analogical as traditionally set forth. Analogy requires a model of proportionality as its basis, and the claims of the traditional argument require a disproportionality between self-knowledge and knowledge of others for their problematic starting point. Nor is it serviceable as a basis for knowledge, since analogies provide ways of understanding rather matters of information. Nor is it appropriately concerned with minds, since the concern seems to be with understanding the other as a person rather than with knowing his mind. Once these reconstructions have been made, the "analogical argument" can lead to enriching one's understanding of himself and others; but it can do so not only because the analogical argument is not an argument, but also because the problem of other minds is not a problem.

AUSTRALASIAN JOURNAL OF PHILOSOPHY
Vol. LI, No. 3, December 1973

Mental Events: Are There Any?, BRANDON TAYLOR

It is a common assumption in discussions of the mind-body problem that there are such things as mental events, and that these comprise noticings, perceivings, assumings, rememberings and the like. An examination of certain grammatical features of "mental language" suggests that this ontological assumption needs to be questioned. The relevant grammatical features include different styles of noun-phrase formation, noun-phrase modification, and some patterns of pronominal usage. The paper ends with a proposal about Nagel's view that a person's possessing a mental attribute can be identified with the same person's possessing a certain physical attribute. The best interpretation of phrases which apparently serve to identify a subject's possessing an attribute, as well as many of the phrases usually thought appropriate for specifying mental events is, in the grammarians' terminology, *factive*. But a fact is not an event.

Proper Names and Their Distinctive Sense, B. R. MILLER

The thesis proposed is that each proper name not only has sense, but one peculiarly its own. In contrasting proper names with pointers and labels, it is shown that their sense is not their reference, but is to refer to what is or was their bearer. The difference in sense between proper names and descriptions or predicates is then based on the distinction between uniqueness of reference and uniqueness of predicability. Proper names refer to a determinate individual, whereas the best the latter can do is to be true of exactly (though any) one individual. Thus the sense of a proper name cannot be captured by any description (precise or imprecise) nor by any predicate (even a unique one).

Don't Care Was Made to Care, ROSS BRADY and RICHARD ROUTLEY

The authors criticize some arguments which have recently been put forward against three-valued significance logic. There are six sections dealing with R. J. Haack's "No Need for Nonsense," *A. J. P.*, Vol. 49 (1971), and there is one section dealing with E. Erwin's *The Concept of Meaninglessness* (Johns Hopkins, 1970). Haack does not care which of the two values, truth or falsity, is given to sentences like "The number 7 dislikes dancing." The authors show that there is philosophical interest in which value assignment is made to such sentences and that the flexibility of the value assignment leads to logical inconsistencies. The authors go on to show that all of Haack's arguments against three-valued significance logic are fallacious. They also show that the central argument of Erwin's book, i.e., the argument establishing that the "so-called meaningless statements" are false, is question-begging.

The Attributive Theory of Mind, JOHN BRICKE

The author examines two arguments, presented in D. M. Armstrong's *A Materialist Theory of the Mind*, against the Attribute Theory (Double Aspect Theory) of the mind. Armstrong argues that the Attribute Theory fails to satisfy two necessary conditions for an adequate theory of mind: it fails to allow the logical possibility of disembodied minds; and it fails to be a scientifically plausible interactionist theory. Granting Armstrong's conditions, the author argues that Armstrong has failed to show that the Attribute Theory cannot satisfy each condition. In defending the Attribute Theory against the two objections, the author discusses the logical relations between alternative versions of the theory, and indicates several interesting (possible) features of the theory.

INTERNATIONAL PHILOSOPHICAL QUARTERLY
Vol. XIII, No. 4, December 1973

Nishida Kitaro: Nothingness as the Negative Space of Experiential Immediacy, DAVID DILWORTH

Linguistic Phenomenology, JERRY H. GILL

The concern is to follow up on J. L. Austin's characterization of ordinary language philosophy as "linguistic phenomenology" and Wittgenstein's notion of bedrock "truths" which, while they cannot be said, "show themselves." An effort is made to view language as a *mediator* of the structure of reality, and examples are offered of contexts within which such bedrock concepts as "meaning," "persons," and "induction" are allowed to disclose themselves. It is submitted that this approach to doing philosophy is more fruitful than either seeking to focus reality directly or merely analyzing linguistic usage.

Happiness: The Role of Non-Hedonistic Criteria in Its Evaluation,
    IRWIN GOLDSTEIN

We should avoid the error of thinking of happiness as a *non-normative,* purely "descriptive" concept on a par with a psychologist's concept of pleasure. Non-hedonic evaluative criteria are also regularly employed in judging the happiness of people. Often, in judging whether or not someone is happy, we not only consider how he feels, but we also evaluate the conditions under which he came to feel the way he does. In the essay, I discuss what it means to be "really happy" or to find "true happiness." The various *grounds* one might have for affirming or denying that someone is *really* or *truly* happy are canvassed. Highly evaluative notions such as "deep happiness" or "a higher happiness" are also discussed. In an important sense, "happiness" suggests an ideal or consummate mental state less often realized than sought.

Is Woman a Question?, MARYELLEN MACGUIGAN

This paper argues that Western philosophy tends to identify the human being with the male human being. Thus women are defined as problematical or imperfect human beings. Masculine bias is found in contemporary as well as past thinkers, even in some who attempt a positive revaluation of woman. Close study of four writers—Ortega, Karl Stern, Buytendijk, Julian Marias—shows that they implicitly take man as the standard of the human and perceive woman as deviant. This bias reflects the assumptions of the male-centered societies in which philosophy originated and has been carried on. Such conceptions of the human being are not adequate for a philosophical reconsideration of women's rights and social role. They need to be critically rethought.

God in the Philosophy of W. E. Hocking: A Centenary Memoriam:
    1873–1973, BARBARA A. MACKINNON

In order to give some much-deserved recognition to William Ernest Hocking in this year of the centenary of his birth, this paper presents an analysis of his philosophy of God, his most central philosophical concern. It presents Hocking's answers to the questions of how we know God, what we know in knowing God, and what role God plays in human life. The paper describes Hocking's attempt to argue that a world mind which possesses both personal and impersonal characteristics can be found by analyzing our most elementary or nuclear experience of the obligation to know things as they are. While containing both scientific and mystical elements, Hocking's philosophy of God is also pragmatic in that he believes such knowledge makes a difference in human affairs. The paper then summarizes his position on the difference it makes in the practices of science, law, and art.

JOURNAL OF THE HISTORY OF PHILOSOPHY
Vol. XII, No. 1, January 1974

Aristotle and Kierkegaard's Existential Ethics, GEORGE J. STACK

The primary aim of this essay is to indicate the relationship between Aristotle's metaphysical language and the conception of the ethical development of the individual in Kierkegaard's philosophical anthropology. Specifically, I argue that the central existential categories which Kierkegaard used to describe the "becoming" of the self are rooted in Aristotle's thought, even though they are modified in such a way as to apply exclusively to human existence. Analogies between Aristotle's understanding of choice and Kierkegaard's "dialectic of choice" (as well as between the former and the latter's conception of potentiality) are presented in order to provide evidence for the influence of Aristotle on Kierkegaard and, incidentally, to dispel the claim that Kierkegaard defends a notion of irrational choice. *En passant,* references are made to the meaning of "absolute choice," the concept of subjective teleology, and the existential category of repetition.

Aristotle's Conception of the Spartan Constitution, ROGER A. DE LAIX

The old arguments concerning Aristotle's empirical or factual approach to history in the *Politics* and the fragments of the 158 Aristotelian *politeiai* should be supplemented or revised through fresh analyses of his treatment of limited, specific themes. The present paper offers an analysis of Aristotle's conception of the Spartan constitution in the *Politics* and the *Lakedaimonion Politeia*. From this examination it is concluded that Books II, VII, and VIII of the *Politics* represent a later, more empirical stage in Aristotle's thinking concerning the Spartan system, and Books IV and V an earlier, more theoretical one. Book V and the *Lakedaimonion Politeia* reflect an intermediate fact-gathering stage in his research. These several stages are particularly clear in Aristotle's change from an approbative to a disapproving view of Spartan institutions, wherein the earlier view of an admirable Lycurgan constitution contrasts with a later critical attitude toward the mistakes of the Spartan "nomothetes."

Harmony and the Heptaplomeres of Jean Bodin,
MARION DANIELS KUNTZ

Bodin's concept of harmony in the *Heptaplomeres* has its grounding in the harmony of nature which serves as an exemplar for man. Coronaeus' home in Venice, the setting of the dialogue, is a harmonius microcosm. Here men of different faiths discuss, with contrasting opinions, the harmony of nature as reflected in numbers, music, and religion. This harmony of nature is based on multiplicity, the aspect of creation in the world. Only Divinity is apart from every multiplicity. Bodin calls upon the harmony of nature to demonstrate that just as there is need for multiplicity in nature, so also is there need for multiplicity in religion and therefore toleration of all religions. Until the harmony of nature, based on multiplicity and diversity, becomes an exemplar for earthly harmony, the lives of men may of necessity reveal contradictions if the state, society, or religion allows for no multiplicity (*concordia discors*), no blending of opposites, no dissonant sounds.

The Inherence Pattern and Descartes' Ideas, THOMAS M. LENNON

The paper shows that sense can be made of Descartes' *ideas,* and the *esse formale–esse objectiva* distinction he draws in discussing them, in terms of the Aristotelian-scholastic account of intentionality. On this interpretation, Descartes' account is relieved of the difficulties raised, for example, by Anthony Kenny ("Descartes of Ideas," *Descartes,* ed. Willis Doney).

Arnauld's Alleged Representationalism, MONTE COOK

Is Arnauld a representationalist? The heated dispute early in this century testifies to the question's difficulty. On one side, John Laird and Morris Ginsberg champion Arnauld for seeing the vulnerability of representationalism and replacing it with something better. On the other side, A. O. Lovejoy and R. W. Church set out to dispel the illusion that Arnauld is anything but a representationalist. I argue that the dispute is spurious, that different uses of the word "representationalism" and a difference in emphasis conceal their fundamental agreement. More important, however, I argue that both sides of the dispute misinterpret Arnauld's theory of perception: both sides fail to appreciate Arnauld's identification of ideas with acts of perception.

Zeno of Elea and Bergson's Neglected Thesis, CONNOR J. CHAMBERS

Careful analysis of the argument of Henri Bergson's secondary doctoral thesis, *Quid Aristoteles de loco senserit* (1889), reveals two important points. 1) The neglected thesis is not at all a defense of Bergson's professed Kantian theory of spatial "realism," for Kant's solution to the problem of spatial extension is shown to be no more successful than the criticized theory of Aristotle. 2) Bergson's debts to Zeno of Elea were two: (a) not only substantive, at least in the negative sense that Zeno's spatial "sophisms" concerning the physical world would remain for Bergson very "respectable difficulties" for years to come; (b) but primarily and positively dialectical, even eristical, in 1889. Taking his theses before a predominantly Kantian faculty, Bergson cleverly illustrated in the Latin work his mastery of Zeno's classical method of academic protest. And illustrated the weaknesses of Kant's theory of space no less than Aristotle's. Which is not to proclaim that college can be fun; yet neither to gainsay that gamesmanship is not the exclusive prerogative of the dissertation committee in its ritual assault upon the classical dialectical thesis of a clever student.

Heidegger on Logic: A Genetic Study of His Thought, THOMAS A. FAY

From the beginning of his career Heidegger has written much on logic. Since his inaugural lecture at Freiburg in 1929 in which he delivered his most celebrated salvo against logic, he has frequently been portrayed as an anti-logician, a classic example of the obscurity resultant upon a rejection of the discipline of logic, a champion of the irrational, and a variety of similar things. Because many of Heidegger's statements on logic are polemical in tone, there has been no little misunderstanding of his position. Frequently the position which is attacked as Heidegger's is a barely recognizable caricature of it. Our purpose in this study is, therefore, twofold. First, we attempt to determine what Heidegger's position actually is by tracing the development of his thought from his first article of 1912 on logic through to his latest works. Secondly, we attempt to see what role, if any, logic would play in what Heidegger calls "authentic thought."

Which Worlds Could God Have Created?, ALVIN C. PLANTINGA

Causation, DAVID LEWIS

A counterfactual analysis of causal dependence and causation (among events in a deterministic world) is stated, defended, and compared with a certain regularity analysis of causation.

Vol. LXX, No. 18, October 25, 1973

Givenness and Explanatory Coherence, WILFRED SELLARS

The Logic of Natural Language, BARBARA H. PARTEE

Claim to Moral Adequacy of a Highest State of Moral Judgment, LAWRENCE KOHLBERG

Vol. LXX, No. 19, November 8, 1973

Mathematical Truth, PAUL BENACERAF

    Two quite distinct kinds of concerns have separately motivated accounts of the nature of mathematical truth: 1) the concern for having a homogeneous semantical theory in which semantics for the propositions of mathematics parallel the semantics for the rest of the language; and 2) the concern that the account of mathematical truth mesh with a reasonable epistemology. It is my thesis that almost all accounts of the concept of mathematical truth can be identified with serving one or another of these masters at the expense of the other. Since I believe further that both concerns must be met by any adequate account, I am dissatisfied with any package of semantics and epistemology that purports to account for truth and knowledge both within and without mathematics. For, as I suggest, accounts of truth that treat mathematical and non-mathematical discourse in relevantly similar ways do so at the cost of leaving it unintelligible how we can have any mathematical knowledge whatsoever; whereas those accounts which attribute to mathematical propositions the kinds of truth conditions we can clearly know to obtain, do so at the expense of failing to connect these conditions with any analysis of the sentences which shows how the assigned conditions are conditions of their *truth*.

Meaning and Reference, HILARY PUTNAM

The Cow on the Roof, PAUL ZIFF

    The title is an archetypal image serving to describe and explicate the present state of foundations in semantics. It is the allegorical exemplar of researchers impressed with the negative thesis of Tarski's *Der Wahrheitsbegriff in den formalisierten Sprachen,* viz.: if the order of the metalanguage is at most equal to that of the language itself a formally correct and materially adequate definition of true sentence cannot be constructed. Such researchers opt for Tarski's definition of truth as the model for a definition of meaning in a natural language. On such a basis it is then a simple matter to establish the correlative negative thesis that in a natural language it is impossible to construct a formally correct and materially adequate definition of meaningful expression in a natural language. Semantic research is thereby relieved of the need for arduous linguistic studies.

Woltenstorff and Bradley on Ontology, EDWIN B. ALLAIRE

Matter, VERE C. CHAPPELL

THE MONIST
Vol. LVIII, No. 1, January 1974

Justice Is Reasonableness: Aquinas on Human Law and Morality,
   JAMES F. ROSS

This paper is a reinterpretation of Aquinas' natural law theory. The notion that human law is "derived" from natural law has been misunderstood to mean something like "inference" or "logical entailment," that cannot be supported by Aquinas' text. The principles of natural law have themselves been misunderstood because the principles operate as *policies,* rather than universal truths, and are not to be applied by universal instantiation as one applies simple universal laws. Aquinas' surprising doctrine that all but the first principles of natural law have exceptions which cannot be eliminated by adding qualifying phrases to the statement of the principles is the key argument in support of the "policy" interpretation of natural law. Lastly, it is argued that Aquinas' fundamental test for the justice of a law is whether or not the law is *reasonable* (a) in relation to a legitimate public objective, (b) as a means for achieving such an objective, (c) was enacted in a reasonable way and (d) has reasonable effects upon the society.

Is Thomas Aquinas a Natural Law Ethicist?, VERNON J. BOURKE

Although natural law plays a part in the ethics of Thomas Aquinas, it is not its central feature. One view of natural law stresses the Will of God as Legislator, while a second emphasizes the origin of ethical judgments in man's own thinking on his natural experiences. The second characterizes the thought of Aquinas. What is right in nature (*ius naturae*) is more important than any law of nature (*lex naturae*). The famous "Treatise on Law" in the *Summa Theologiae* depends on the metaphysics and anthropology that precede it. Before the *Summa contra Gentiles* (Book III) Thomas' view of law is undeveloped. It comes to maturity in the later works with the emphasis on human reasoning about man's relations to his environment. This theory of right reason makes Aquinas' ethics distinctive.

Transcendental Thomism, J. DONCEEL, S.J.

This paper tries to explain the origin and some of the main tenets of what is sometimes called *Transcendental Thomism*. Its founder, Joseph Maréchal, S.J., trying to understand Kant and to refute his agnosticism, found many valuable elements in Kant's philosophy. They made him read Aquinas with new eyes and helped him discover in St. Thomas the germs of our modern anthropocentric approach in philosophy. The paper explains the ensuing new interpretation, within a thomistic framework, of the relation between senses and understanding, between understanding and reason (intellect), of abstraction and the formation of concepts. The paper insists on the manner in which Transcendental Thomism interprets metaphysics and defends it against its opponents (retorsion). Finally it tells us why this new philosophical trend feels entitled to call itself Thomism.

Aquinas and the Five Ways, JOSEPH OWENS

According to Aquinas' *Summa of Theology*, one proves in five ways that God exists. Those "five ways" are arguments taken from sources in which they did not reach the conclusion claimed for them by Aquinas, and elsewhere "ways" up to eleven in number are used by him without concern for alignment with that fivefold schema. This "enigma of the five ways" results from the use of a metaphysical, not physical or cosmological, framework developed early by Aquinas. Arguments, such as the Anselmian, that do not fit into that framework are rejected. Arguments that do are accepted, but their number and arrangement varies widely with the convenience of the moment. This seems to have been Aquinas' own attitude towards the "five ways."

St. Thomas Aquinas in Maimonidian Scholarship, JACOB I. DIENSTAG

Between Immortality and Death: Some Further Reflections on the Summa Contra Gentiles, ANTON C. PEGIS

The present discussion continues two previous articles on the interpretation of the *Summa Contra Gentiles* of St. Thomas, and especially the role of Aristotelianism in the theology of the relations between soul and body. Grant that the unity of man, both in his existence and in his essence, can be defined and explained only if the soul (an intellectual substance) is the substantial form of the body (a corruptible organism), and the problem of the relations between immortality and death poses itself in an acute form. Is death natural to the human composite? Is immortality? St. Thomas cannot divide man between immortality and death. Rather he argues that, though death is natural for the body, it is not natural for the human person in his unity; and though only the soul in man is by nature immortal, still, given the unity of the human person, immortality does not necessarily involve the separation of the soul from the body. St. Thomas describes that separation as against the nature of the soul.

Is Prudence Love?, CHARLES J. O'NEIL

The question is directed to Aquinas. The author says Aquinas answers: "Love moves within prudential actuality." Metaphysical exegesis of that text must ask how Aquinas could put the source of the act of a virtue of intellect in a habit of the will. The author first shows that the actualities of the intellect's virtue of faith are love's actualities for Aquinas. And in the same way one sees with him the prudential in actuality in every discernible act of courage, of poise, of the use of synderesis, of varying acts of justice. The texts of Aquinas show this: whenever he speaks as one who sees the courageous man, the poised man, the just man as *being-actually* such, he also speaks as one seeing a man in whom there is at once the actuality of prudence and the productive actuality of love.

Ghazali and Aquinas on Causation, ROBERT E. A. SHANAB

The Islamic Medieval Philosopher al-Ghazālī, known to the latins as Algazel (1059–1111), was influential in the shaping of the intellectual philosophic movements in the thirteenth century. Granted that Ghazali's philosophic ideas did not assume a definite pattern in the philosophic literature as was the case with Avicenna and Averroës, still a careful study of Ghazali's works will reveal how profound and widespread his influence was on western Medieval scholars. A case in point is the influence of Ghazali on Saint Thomas Aquinas who studied the works of the Islamic philosophers, especially Ghazali's, at the University of Naples. The aim of this paper, however, does not consist in delineating the many points of encounter between the two philosophers. Rather it deals primarily with their discussion of the principle of causality, and how the works of Ghazali played an important role in the shaping of the philosophic ideas of Aquinas.

Aquinas' Definition of Good, RONALD DUSKA

In Q. 21 of the *De Veritate*, Thomas Aquinas states that the essence of good is to be perfective after the manner of an end. This claim, when viewed from an ethical-theoretical perspective, shows that to call something "good" requires that the thing be (a) perfective of another and (b) desired and/or enjoyed. It is argued that such requirements yield a negative empirical test for determining good which speaks to the contemporary debate between naturalists and non-naturalists; and that because Thomistic exegesis usually concentrates on the content of Aquinas' ethics, it overlooks a significant formal test for deciding whether something is good or not.

"The Five Ways"—Proofs of God's Existence?, L. C. VELECKY

The thesis that *ST* I, 2, 3 is to be interpreted as providing "proofs of God's existence" is criticized because the arguments against it seem stronger than those for it. The positive suggestion made is that Aquinas provides here *"ratio huius nominis 'deus' "* and tries to integrate the proposition "God exists" into the "universe" of human discourse. This proposition is linked to the logical conclusion in the form "There is something to which a given predicate applies" by means of the factual statement "And this is what everybody calls 'God'." This suggests that the five arguments assume the legitimacy of the distinction Creator–creature which all attempted disproofs of God's existence assume to be illegitimate. Since neither assumption can be justified by any prior purely rational principle it seems that no one can either prove or disprove by purely rational arguments that God exists.

Saint Thomas Aquinas' Theory of Universals, RALPH W. CLARK

I contrast the traditional interpretation of St. Thomas' theory of universals (that universals exist in things) with the interpretation which finds its fullest development in journal articles by Joseph Owens (that universals exist only in the intellect, while common natures in things are neither one nor many). I claim that neither interpretation is correct, that a common nature in things is, according to Thomas, not a universal but nevertheless is one in those things. It is Thomas' position, I claim, that a universal is "more one" than is a common nature in things. I defend these claims by producing textual support for them and by showing that Thomas is committed to them as a consequence of his distinction between essence and existence and some claims which he makes regarding the nature of concepts.

THE NEW SCHOLASTICISM
Vol. XLVII, No. 1, Winter 1974

The Equivocal Use of the Word "Analogical," MORTIMER J. ADLER

This article seeks to throw light on what is involved in saying that there is a middle ground between a univocal and an equivocal use of words without using the word "analogical" to signify that middle ground. We thereby avoid an equivocal use of the word "analogical" to signify (i) what, on the one hand, is only a mode of equivocation and (ii) what, on the other hand, is a mode of naming which is neither univocal nor equivocal. I propose to do this in three steps. In the first, I will deal solely with Aristotle's doctrine of the univocal and the equivocal, and with his account of diverse modes of equivocation. In the second, I will attempt to show that the modes of the analogical discussed by Aquinas are all reducible to Aristotle's modes of equivocation. In the third, I will draw certain conclusions from my argument that the middle ground between the univocal and the equivocal should neither be called a mode of equivocal speech nor a mode of analogical speech.

What Cannot Be Said in St. Thomas' Essence-Existence Doctrine,
W. NORRIS CLARKE

The essence-existence doctrine of Aquinas contains at least three elements which cannot be said directly in language, though they can be recognized by the mind in an act of language-transcending insight: 1) the name of God as Subsistent Act of Existence, transcending the ultimate linguistic distinction of subject and verb; 2) the being of participated existence as somehow truly in its participants and unifying them, yet neither existing nor one in itself save in the mind; hence somehow transcending the distinction between one and many—intelligible finally only in the light of the unity of the original creative intentional act of the source as willing to share its perfection with many; and 3) the being of essence as limit in the indissoluble yet irreducible unity-duality-whole that is limited being, limited perfection. In an important sense, then, what cannot be said in the Thomistic essence-existence doctrine is almost, if not equally, as important as what can be said.

Soul As Agent in Aquinas, JOSEPH OWENS

Aquinas accepts wholeheartedly the Aristotelian tenet that the soul is the form of the body. Unlike Aristotle, however, he is thinking in an existential framework that requires the human soul to be an existent and an agent in itself. Through precisive abstraction the soul is regarded as the agent in the intellective and voluntary order, the body as the agent in the other activities. Likewise for Aquinas the soul continues to be an agent in its perpetual existence after bodily death. But this separated activity is difficult to demonstrate on purely philosophical grounds in the Thomistic framework. Though accepted as a fact it seems to remain metaphysically an aporia and a mystery.

Kant and Aquinas, HENRY VEATCH

Kant rejects the idea that moral principles can be rationally justified if such principles involve human desire. For to act from desire or inclination is to act from ordinary natural causes and no reasons can be given in justification of natural events. Aquinas distinguishes two kinds of appetites: sensory and intellectual. Kant rejects sensory appetites as being morally irrelevant to one's actions. But Kant never recognizes intellectual appetites, and it is just these appetites which Aquinas contends fulfill the demand that they be rationally justifiable. Objects of our intellectual desire determine our desire or will, not in the manner of mere necessary causes, but rather in the manner of good and sufficient reasons. It is by means of this distinction that Aquinas can provide for the rational justification of moral principles and yet hold that one's desires are relevant to how one ought or ought not to act.

Thomas' Evaluation of Plato and Aristotle, JAMES A. WEISHEIPL

Thomas Aquinas knew a great deal about the philosophy of Plato, mainly through secondary sources. This paper considers five basic issues of philosophy on which Thomas explicitly stated his views of the teaching of Plato and Aristotle: 1) the nature of man; 2) the process of knowledge; 3) subsistent Ideas; 4) separated substances; and 5) the existence and nature of God. In some minor instances Thomas preferred the view of Plato to that of Aristotle, but on the major issues, Thomas decidedly preferred the views of Aristotle.

Were the Oxford Condemnations of 1277 Directed Against Aquinas?, LELAND EDWARD WILSHIRE

It is a commonly held assumption that the Oxford Condemnations of 1277 were directed against Aquinas and Unity of Form. This interpretation cannot be supported by Archbishop Kilwardby's speech, by the historical context or by the subsequent correspondence. The articles deal with issues that predate Thomistic formulations. The historical context shows no conflict with Thomism and Kilwardby's letter to Peter of Conflans, a young disciple of Thomas, the best evidence on the speech, specifically states that "Unity of Form" was not condemned at Oxford and that the whole issue was only then beginning to dawn upon the aged Archbishop.

NOUS
Vol. VII, No. 4, November 1973

Criteria and Necessity, RICHARD RORTY

The notion of "criterion" has become that of something half-way between a condition entailed by a definition and an empirical generalization. But suspicion about the analytic-synthetic distinction should make one suspicious about the phrase "entailed by a definition" and thus about the need for such half-way houses. As used by writers like Malcolm and Strawson, a criterion for X's seems to be simply a truth about X's known on some basis other than empirical correlations of X's with other things. But there is no need to postulate a special source (e.g., "meanings of terms") for such knowledge. Nor do we need such an epistemological theory to bring out the force of Wittgenstein's claim that inner processes require outward criteria.

Possible Individuals, Trans-World Identity, and Quantified Modal Logic, NICHOLAS RESCHER and ZANE PARKS

Why Settle for Anything Less Than Good Old-Fashioned Aristotelian Essentialism, BARUCH BRODY

In this paper, I defend a theory of essentialism based upon Aristotle's distinction between mere alterations and substantial changes. A substance has a property accidentally if its losing it would be a mere alteration while it has it essentially if its losing it would mean its going out of existence. I defend this theory of essentialism against the various recent criticisms of essentialism, especially the one based upon the problem of identity through possible worlds. I show that it offers a more intuitively satisfactory account of what are essential properties than the accounts offered by Plantinga and Kaplan. Finally, I discuss the problems involved in constructing a theory of natural kinds based upon this theory of essentialism.

Son of Grue: Simplicity vs Entrenchment, KENNETH FRIEDMAN

It is claimed not only that entrenchment is too superficial a feature to give a satisfactory account of projectibility, but also that there are counterexamples to the account of projectibility in terms of entrenchment. Examples are given in which the better-entrenched predicates are less projectible. It is argued that according to an earlier account of simplicity (*BJPS, 23*), the simpler a class for a person the more projectible that class for that person. So, for example, were a person to directly perceive grue, rather than green, the class of grue things would be simpler, and so more projectible for him. It is further shown, given weak and reasonable assumptions, that people associate simpler classes with earlier learned and more familiar predicates. Finally, it is argued that the use of complex Goodmanesque predicates would tend to evolve to the use of simpler predicates.

Ambiguity and the Truth Definition, KATHRYN PARSONS

In this paper I argue that the presence of ambiguous terms raises a serious question about Quine's claims that a Tarski definition can be given for a regimented language, and about his claim that the Tarski paradigm adequately clarified the notion of truth in ordinary language. It also raises a difficulty for Davidson's programme of using a Tarski-like analysis to give a theory of meaning for natural languages. In the end, a question is raised about the Tarski "material adequacy condition" itself.

The Possible In the Actual, ADAM MORTON

I try to show that one can give a semantical analysis of modal idioms which does not involve construing any expressions of the object language as referring to possibilia. I first say in what sense the standard modal semantics do construe modal idioms as referring to possibilia, and then I describe an alternative account which is better behaved in this respect. The central idea is that by referring to additional properties of actual individuals one can define structures, "partial worlds," which can serve the purposes for which possible worlds, structures involving additional individuals, are usually used.

Opacity and the Ought-to-Be, LOU GOBLE

THE PHILOSOPHICAL QUARTERLY
Vol. XXIV, No. 94, January 1974

Categories in Natural Language, A. G. ELGOOD

I use "category allocation" to mean either the allocation of a and b to the same category (Sab) or the allocation of a and b to different categories (Dab). "Category allocation rule" I use to mean a conditional or biconditional statement which has a category allocation on one side and on the other a statement about the sense-values of sentences which conjoin a and b with at least one other term or with all other terms. Evidently some philosophers sometimes use arguments which include category allocation rules in this sense and I give examples of these. In this paper it is my purpose to show that such arguments cannot lead to philosophical enlightenment, and this I do both by criticizing particular category allocation rules offered by philosophers and by employing arguments against *any* such rules.

For Ever?, R. F. HOLLAND

This article is an argument in favor of the position that there is for our theoretical understanding no such thing as the absolute or eternal "for ever" of prolonged existence.

CURRENT PERIODICAL ARTICLES 645

Logic of Terms, G. D. DUTHIE

This article introduces an exceptionally compact notion for those parts of the classical calculus of one-place predicates which are most used in actual reasoning. It is based on the form of the "involution" and is in other respects also closely analogous to the author's "Intensional Propositional Logic" (*Philosophical Quarterly*, Jan. 1970), many of whose theses are applied in the system of the present paper. All wffs in the system have a logically equivalent canonical form, and a decision procedure is provided. The economy of the notation is illustrated by a succinct exposition of the traditional syllogistic. Foundations are laid for a theory of enthymemes. By the formation rules tautological and trivial formulae ("all A's are either A or B") cannot occur.

Reasons for Action, SCOTT MEIKLE

The Truth About Fictional Entities, H. GENE BLOCKER

The usual Strawsonian account of referring will not do for fictional entities. The problem is that we still do not have a sufficiently clear notion of ordinary referring, and the root of this problem is that, despite the important distinction drawn between denoting and referring, referring is still perceived in terms of a paradigm relation of a description to an existing thing. But that relation, it is argued, is preceded by the more fundamental referential relation of thought to an object of thought, whether real *or* imaginary. The conclusion reached is that fictional reference is an institutionalized partial use of ordinary referring, parasitic on it though leading a legitimate life of its own, and sharing with ordinary referring the crux of reference, our understanding partial, generalized descriptive expressions to be descriptions of complete, concrete objects and organizing different descriptions round the same objects.

THE PHILOSOPHICAL REVIEW
Vol. LXXXII, No. 2, October 1973

The Hypothetical Imperative, THOMAS E. HILL, JR.

Particular moral judgments, Kant believed, presuppose a general principle of rationality, the Categorical Imperative; particular prudential judgments, he implied, presuppose another principle of rationality, the Hypothetical Imperative. This principle, which has been comparatively neglected, has strikingly many of the features Kant attributed to the Categorical Imperative. The questions, then, are these: What exactly does the Hypothetical Imperative prescribe? In what sense is it an imperative, and in what sense hypothetical? Why did Kant accord such a special sublimity to the Categorical Imperative when the Hypothetical Imperative is a rational principle similar in so many respects? Among the conclusions: although neither the Hypothetical Imperative nor the Categorical Imperative expresses merely a prima facie requirement of reason, it is always possible in principle to satisfy both principles completely; and there is a coherent explanation for Kant's view that, though immoral acts are not excusably unfree, the morally conscientious person is the most fully free.

A Reasonable Self-Predication Premise for the Third Man Argument, SANDRA PETERSON

Essentialism, Possible Worlds, and Propositional Attitudes, GAIL C. STONE

PHILOSOPHY
Vol. XLVIII, No. 186, October 1973

Under What Net?, WILLIAM K. FRANKENA

In *Morality and Art* Mrs. Foot (a) defines morality as aiming at a certain object or end, viz., "removing particular dangers and securing certain benefits," and (b) infers "that some things do and some do not count as objections to a line of conduct from a moral point of view." In this article her non-formalist conception of morality is accepted, but (a) and (b) are criticized, and a non-poietic view of morality, which takes it as a kind of doing rather than as a kind of making (in Aristotle's sense), is proposed.

An Irrelevance of Omnipotence, P. T. GEACH

The problem of evil is often stated in relation to the doctrine that God is omnipotent, i.e., able to do everything. Opponents of Christianity often argue that omnipotence is an incoherent concept: but if so, there just is no problem how a world containing such evils as there are could have been created by an *omnipotent* God. The question of giving an acceptable sense to "able to do everything" is in any case a distraction. For Christians are certainly committed to the view that God is *almighty,* i.e., that his will cannot be frustrated and no outside thing can restrict his action; and this means, as McTaggart pointed out, that God's antecedent and consequent will cannot be distinguished as is traditionally done in theodicies; moreover, as Aquinas already said, an almighty God cannot be caused to will means by the prospect of an end. Controversialists on both sides would be well advised to concentrate on the difficulties of almightiness.

Punishment and Remorse, JENNY TEICHMAN

Understanding a Primitive Society, H. O. MOUNCE

Common Sense Propositions, A. C. EWING

Vol. XLIX, No. 187, January 1974

Mowgli in Babel, GILBERT RYLE

Liberalism, H. J. McCLOSKEY

In the first part of this article the commitment of liberalism to coercive and non-coercive action based on value judgments is examined by reference to the concepts and ideals of liberty, its operative view concerning the role of government, and its concern for morality and moral values. In the second part, the value commitment of liberalism is illustrated by reference to its concern with fostering self development and the conditions thereof, education, especially moral education, its involvement in no-compromise issues such as relate to racialism and abortion. Finally the value judgments involved in taking a stand in respect of protest, disobedience and rights of conscientious objection are noted.

Religious and Secular Statements, D. H. MELLOR

Religious (specifically Christian) explanations of events are often supposed to conflict, in a logically obscure way, with secular (especially scientific) ones. This paper sets out to clarify and dispose of the issue by exposing the presuppositions of the conflict and showing that they do not obtain. It urges that religious statements differ from secular ones in not being backed up by an appropriate theory of religious perception that could resolve a conflict with secular statements in favor of the former. It follows that secular ontology is immune to erosion by religious skepticism, but not vice versa. There is no harm, however, in Christians admitting their exposure to possible empirical refutation in exchange for actual empirical support. Given that, no further conflict between secular and religious explanations need arise; given that they are neither logically nor causally incompatible. It is, in particular, perverse to insist nonetheless that there is a conflict and then resort to obscure logical relations like "complementarity" to resolve it.

Austin's Mistake about 'Real', D. J. C. ANGLUIN

There are characteristic ways, for Austin, in which a thing might fail to be what it seems, so that the sense of the claim that it is a real S (e.g., duck) is given by the ways excluded, where each way answers to a criteria-governed expression (e.g., 'decoy'). Thus 'real' has different "uses" and, in each use, in criteria-governed; further, the ability to make such reality claims is apparently acquired use by use. I argue, however, that it is not set uses which are learnt, but the purposes one has in making such claims, and that this knowledge accounts not only for the learning of the uses, but also for how they can come to modified. This both explains phenomena which Austin cannot explain, and disallows his argument against traditional metaphysics.

Body and Soul in Aristotle, RICHARD SORABJI

Aristotle has been called a materialist and a Cartesian. He has also been assimilated to Brentano and to Ryle. In fact, however, he has a unique position of his own on the relation between body and soul. When we see what his position is, we notice that it has interesting implications for the traditional body-soul problems.

PHILOSOPHY AND PHENOMENOLOGICAL RESEARCH
Vol. XXXIV, No. 2, December 1973

On Unconscious Emotions, MICHAEL FOX

This article is a critique of the dispositional analysis of unconscious emotions, favored by many recent critics of psychoanalysis, with special reference to that offered by Harvey Mullane in "Unconscious Emotions" (*Theoria*, 1965). It is argued that what Freud designated as "unconscious emotions" are *not* the objectionable "unfelt feelings" which dispositional theorists suppose, but rather, *presently experienced* states, which must be conceptualized as self-deceivingly misdescribed or misrepresented by the patient to himself. Support for this view is found both in Freud's own words and in an extensive analysis of Mullane's examples. Finally, it is maintained that only in the case of hysterical conversion are we (logically) required to resort to a dispositional analysis of psychoanalytic statements about unconscious emotions.

The Canon of Subjectables, DOUGLAS BROWNING

The assumption that it is possible to refer to any singular by means of a subjecting-expression, i.e., an expression which serves to subject a singular to predicative characterization, is shown to be questionable. It is argued that (1) there are other means of making references than by using subjecting-expressions; (2) at least some singulars referred to by these other means "are;" and (3) at least some of these latter cannot also be referred to by means of subjecting-expressions.

A Philosophy of Despair, BERNARD BYKHOVSKI

The contraposition of thought and action is basic to Kierkegaard's world-view. Philosophy, especially Hegel's, is futile. Feeling, not reason is the engine of human existence. The transition from knowledge to action is not a logical one. It can be made only by faith. Self-mastery through suffering for sin is the key to existence. Kierkegaard's criticism of Hegel is the direct opposite of Marx's. Marx's criticism transforms the world, Kierkegaard's rejects it; Marx's way is based on the factual, Kierkegaard's on the subjective and absurd. Marx's way is from theory to political practice; Kierkegaard's from reason to faith. The rational core of Kierkegaard, his call for action and criticism of speculation, is vitiated by his rejection of objectivity and reason. He does not aim at either understanding the world or changing it. His solution to the hopelessness pervasive today is irrationalism. The concept of development is alien to his philosophy; his reply to difficulties is despair. Its outgrowth, existentialism, flourished after the two world wars, when the old tradition had collapsed. But it was incapable of even imagining a solution. Whereas an adequate philosophy orients the rational theory of action, Kierkegaard presents only disorientation in every sphere. His is the philosophy of a doomed society.

Law and Revolution, E. P. PAPANOUTSOS

"Descriptive" and "Revisionary" Metaphysics, DEREK A. McDOUGALL

This paper is a discussion of the distinction between "Descriptive" and "Revisionary" metaphysics as put forth by P. F. Strawson in his book, *Individuals.* Although from one point of view all metaphysics must ultimately be "revisionary" because all theories of this kind go beyond a pre-theoretical level in human experience, the very fact that such a level can be established within a philosophical discussion points to the justification for the descriptive metaphysician's approach. His analysis of the categorical structure of human thinking, unlike those of some extreme revisionary metaphysicians, results in a picture of the world which is at least not at odds with our commonly held notions of ourselves as persons. The notion of a pre-theoretical level in experience having been argued for, it then remains to show why, in the nature of the case, philosophy, as thought about the structure of human thinking, must go beyond it.

Language, Speech, and Speech Acts, EDWARD MacKINNON

There is an unresolved and generally unrecognized conflict between the Wittgensteinian analysis which explicates the meaning of a word through the role it plays in language games and the Austin-Searle analysis which stresses speech acts and the intentionality of the performer as a basis of meaning. To clarify and partially resolve this conflict de Saussure's distinction between language and speech is revived and utilized. Though the two domains are intimately interrelated the problemmatic of meaning analysis assumes a different form in each domain. To bring this out it is necessary to distinguish between the role and rule dependent sense of words in language and the performance dependent meaning of words in speech as well as between denotation in language and reference in speech. The bearing such distinctions have on some broader issues is briefly indicated.

ANNOUNCEMENTS

The John Dewey Foundation and the Center for Dewey Studies, has selected as the theme of its 1974 John Dewey Contest, "The Ethical Theory of John Dewey." The contest is open to matriculated graduate students. All essays must be under 10,000 words and should be received by the Center no later than June 1, 1974. The judges will be Charles Frankel (Columbia University), Lewis E. Hahn (Southern Illinois University at Carbondale), and Charles S. Stevenson (University of Michigan). A final decision and the announcement of two awards, $1,000 and $500, will be announced by September 1, 1974. For further information contact Jo Ann Boydston, Director, Center for Dewey Studies, Southern Illinois University, Carbondale, Illinois 62901.

The first issue of the new *Journal of Chinese Philosophy*, devoted to the study of Chinese philosophy and Chinese thought, appeared in December 1973. Chung-Ying Cheng (University of Hawaii) is the editor; Antonio A. Cua (The Catholic University of America) is associate editor. The first issue contains the following articles: Chung-Ying Cheng, "Statement Concerning the Founding of the *Journal of Chinese Philosophy*;" Thomé H. Fang, "A Chinese Philosophical Glimpse of Man and Nature;" Benjamin Schwartz, "On the Absence of Reductionism in Chinese Thought;" Arthur Danto, "Language and the Tao: Some Reflections on Ineffability;" Antonio S. Cua, "Reasonable Action and Confucian Argumentation;" Chung-Ying Cheng, "On Zen (*Ch'an*) Language and Zen Paradoxes;" and Robert S. Cohen, "The Problem of 19(k)." All manuscripts and editorial correspondence should be addressed to the editor: Professor Chung-Ying Cheng, Journal of Chinese Philosophy, Department of Philosophy, University of Hawaii, Honolulu, Hawaii 96822. For subscriptions write D. Reidel Publishing Company, P.O. Box 17, Dordrecht, Holland.

The recently formed International Society for Neoplatonic Studies invites those interested in Neoplatonic studies to join the Society. The present Executive Committee is composed of John Anton (Emory University), Chairman; J. N. Findlay (Boston University); John Fisher (Temple University); and R. Baine Harris (Old Dominion University), Executive-Secretary and Treasurer. Correspondents for the Society have already been established in Great Britain, Canada, France, Germany, and Greece. Address inquiries to Professor R. Baine Harris, Chairman, Department of Philosophy, Old Dominion University, Norfolk, Virginia 23508.

The Third Hume Conference will be held on the campus of Northern Illinois University, October 25 and 26, 1974. Papers dealing with Hume's philosophy and writings are welcome. Two copies of each paper should be received by August 1, 1974. Correspondence should be addressed to Professor James King, Northern Illinois University, DeKalb, Illinois 60115.

A new journal of interdisciplinary normative studies, entitled *Reason Papers*, has been announced. *Reason Papers* will be edited by Professor Tibor R. Machan (State University of New York at Fredonia). Associate editors include Professors Eric Mack (Eisenhower College), Ralph Raico (State University of New York at Buffalo), Mary Sirridge (University of Massachusetts), John O. Nelson (University of Colorado), John Cody (Northwestern University), and Walter Block (Bernard Baruch College). Manuscripts are welcome on all normative topics. All correspondence, including payment of $3.00 for one copy of issue No. 1, should be sent to T. R. Machan, Editor, *Reason Papers*, Department of Philosophy, State University College, Fredonia, New York 14063.

The Department of Philosophy at Illinois State University is pleased to announce the appointment of Kenton F. Machina (University of California, Los Angeles) and J. Andre Cadieux (Minnesota) as assistant professors.

Professor Milton K. Munitz (formerly of New York University) has joined the Philosophy Department of Baruch College, City University of New York, as Distinguished Professor.

A conference on morality and international violence will be held at Kean College of New Jersey on April 22–24, 1974. Speakers include: John Hospers, Joseph Margolis, Gerald Dworkin and Kurt Baier on Collective Responsibility for Acts of War; Richard Wasserstrom, Hugo Bedau, Marshall Cohen and Martin Golding on Intention and Responsibility in Modern War; Alasdair MacIntyre, H. D. Aiken, P. H. Nowell-Smith and Mary Mothersill on The Limits of the Use of Force in Modern Warfare; R. M. Hare, Kai Nielson, Carl Wellman and Shlomo Avineri on Terrorism; George Schrader, Arthur Danto and Kenneth Mills on the Feasibility of Moral Codes in Modern Warfare; Paul Ziff, John Ladd and David P. Gauthier on Aggressive War and the Limits of National Sovereignty. For further details contact Robert Sitelman, Department of Philosophy, Kean College of New Jersey, Morris Avenue, Union, New Jersey 07083.

The Faculty of Philosophy of the University of Ottawa, in conjunction with scholars from other Canadian Universities, is organizing an International Congress marking the 250th anniversary of Immanuel Kant. The title of this Congress is "Kant in the English-speaking and Continental Traditions" and it will be held at the University of Ottawa, Ottawa, Canada, October 10–14, 1974. Address inquiries to Executive Secretary, Ottawa Kant Congress, Faculty of Philosophy, University of Ottawa, Ottawa, Ontario, K1N 6N5, Canada.

The Department of Psychiatry of the College of Physicians and Surgeons of Columbia University, New York, New York, and the Foundation of Thanatology will co-sponsor a Symposium on Acute Grief and the Funeral at the Columbia-Presbyterian Medical Center on March 29th and

30th, 1974. This conference will take place in Maxwell Hall, 179 Fort Washington Avenue, New York, New York. For information write to Dr. Austin H. Kutscher, 630 West 168th Street, New York, New York 10032.

NECROLOGY

Alan Ross Anderson, Professor of Philosophy at the University of Pittsburgh, died on December 5, 1973. He had previously taught at Dartmouth and at Yale, and had been chairman of the Department of Philosophy at Pittsburgh. He was noted for his work in modal and doentic logic and in the theory of normative concepts. His students and colleagues will remember not only his contributions to philosophy, but also his warmth and humor.

Imre Lakatos died suddenly on February 2, at the age of 51. One of the foremost philosophers of mathematics in his generation, he is also remembered for his contributions to the philosophy of science. Born in Hungary on November 9, 1922, he studied at Debrecen, Budapest, and Moscow. After the Hungarian uprising he escaped to Vienna and with the help of a Rockefeller fellowship, he went to Cambridge to study under Brathwaite and Smiley. He eventually became Professor of Logic at the London School of Economics. From 1964 he was a frequent visitor to the United States.

Horace M. Kallen, philosopher and educator, who was a founder of the New School for Social Research, died February 17. The author of more than 30 books and active in liberal, educational, and Zionist movements, he taught at the New School from 1919 to 1969 serving as dean of its Graduate Faculty of Political Science 1944–1946. Born in Berenstadt, Germany, he was taken to Boston at the age of 5. He later attended Harvard from which he received his B.A. *magna cum laude* in 1903 and his Ph.D. in 1908. An assistant to George Santayana, he subsequently became an editor of some of the work of William James and a collaborator with John Dewey. Among his most recent books are *Liberty, Laughter and Tears* and *What I Believe In and Why—Maybe*.

U. S. POSTAL SERVICE
STATEMENT OF OWNERSHIP, MANAGEMENT AND CIRCULATION
(Act of August 12, 1970: Section 3685. Title 39. United States Code)

SEE INSTRUCTIONS ON PAGE 2 (REVERSE)

1. TITLE OF PUBLICATION: **The Review of Metaphysics**
2. DATE OF FILING: **October 1, 1973**
3. FREQUENCY OF ISSUE: **Quarterly**
4. LOCATION OF KNOWN OFFICE OF PUBLICATION *(Street, city, county, state, ZIP code) (Not printers)*: **The Catholic University of America, Washington, D. C. 20017**
5. LOCATION OF THE HEADQUARTERS OR GENERAL BUSINESS OFFICES OF THE PUBLISHERS *(Not printers)*: **The Catholic University of America, Washington, D. C. 20017**
6. NAMES AND ADDRESSES OF PUBLISHER, EDITOR, AND MANAGING EDITOR

PUBLISHER *(Name and address)*: **Philosophy Education Society, Inc., Catholic University of America, Wash., D.C. 20017**

EDITOR *(Name and address)*: **Jude P. Dougherty, 620 Michigan Avenue, N. E., Catholic University of America, Washington, D. C. 20017**

MANAGER EDITOR *(Name and address)*: **William A. Frank, 620 Michigan Avenue, N. E., Catholic University of America, Washington, D. C. 20017**

7. OWNER *(If owned by a corporation, its name and address must be stated and also immediately thereunder the names and addresses of stockholders owning or holding 1 percent or more of total amount of stock. If not owned by a corporation, the names and addresses of the individual owners must be given. If owned by a partnership or other unincorporated firm, its name and address, as well as that of each individual must be given.)*

NAME	ADDRESS
Philosophy Education Society, Inc. (non-profit corporation)	The Catholic University of America, Washington, D. C. 20017

8. KNOWN BONDHOLDERS, MORTGAGEES, AND OTHER SECURITY HOLDERS OWNING OR HOLDING 1 PERCENT OR MORE OF TOTAL AMOUNT OF BONDS, MORTGAGES OR OTHER SECURITIES *(If there are none, so state)*

NAME	ADDRESS
The First National Bank of Washington	Washington, D. C.

9. FOR OPTIONAL COMPLETION BY PUBLISHERS MAILING AT THE REGULAR RATES *(Section 132.121, Postal Service Manual)*

39 U. S. C. 3626 provides in pertinent part "No person who would have been entitled to mail matter under former section 4359 of this title shall mail such matter at the rates provided under this subsection unless he files annually with the Postal Service a written request for permission to mail matter at such rates."

In accordance with the provisions of this statute I hereby request permission to mail the publication named in Item 1 at the reduced postage rates presently authorized by 39 U. S. C. 3626

(Signature and title of editor, publisher, business manager, or owner)

10. FOR COMPLETION BY NONPROFIT ORGANIZATIONS AUTHORIZED TO MAIL AT SPECIAL RATES *(Section 132.122, Postal Manual)*
(Check one)

The purpose, function, and nonprofit status of this organization and the exempt status for Federal income tax purposes — ☒ Have not changed during preceding 12 months ☐ Have changed during preceding 12 months *(If changed, publisher must submit explanation of change with this statement.)*

11. EXTENT AND NATURE OF CIRCULATION	AVERAGE NO. COPIES EACH ISSUE DURING PRECEDING 12 MONTHS	ACTUAL NUMBER OF COPIES OF SINGLE ISSUE PUBLISHED NEAREST TO FILING DATE
A. TOTAL NO. COPIES PRINTED *(Net Press Run)*	3600	3800
B. PAID CIRCULATION 1. SALES THROUGH DEALERS AND CARRIERS, STREET VENDORS AND COUNTER SALES	0	0
2. MAIL SUBSCRIPTIONS	3400	3400
C. TOTAL PAID CIRCULATION	3400	3400
D. FREE DISTRIBUTION BY MAIL, CARRIER OR OTHER MEANS 1. SAMPLES, COMPLIMENTARY, AND OTHER FREE COPIES	100	100
2. COPIES DISTRIBUTED TO NEWS AGENTS, BUT NOT SOLD	5	5
E. TOTAL DISTRIBUTION *(Sum of C and D)*	3505	3505
F. OFFICE USE, LEFT-OVER, UNACCOUNTED, SPOILED AFTER PRINTING	95	295
G. TOTAL *(Sum of E & F—should equal net press run shown in A)*	3600	3800

I certify that the statements made by me above are correct and complete

(Signature of editor, publisher, business manager, or owner) William A. Frank

PS Form 3526 July 1971

INDIAN PHILOSOPHICAL QUARTERLY

(in New Series)
(Formerly The Philosophical Quarterly)

A quarterly journal of the Pratap Centre of Philosophy, Amalner and the Department of Philosophy, University of Poona, Poona—7.

Edited by S. S. Barlingay

Editorial Board: K. D. Bhattacharya, T. M. P. Manadevan, Shivajivan Bhattacharya, Rajendra Prasad, K. J. Shah, G. Misra, R. C. Pandey, R. K. Tripathi, Daya Krishna, D. P. Chattopadhyaya, G. N. Joshi, D. Y. Deshpande, J. De. Marneffe, R. Sunder Rajan, Dharmendra Kumar, R. C. Gandhi, Y. J. Khwaja, S. A. Shaida.

Volume 1   *New Series*   *Number 2*

Dhirendra Sharma: Phenomenology of Religion and Sri Aurobindo.
S. J. Simmon Decloux: Feuerbach and the Young Marx.
S. A. Shaida: Moore's Evaluation of Sidgwick's Hedonism.
Sushil Kumar Saxena: Laya or Musical Duration in Hindustani Music.
W. S. Barlingay: Dr. Ambedkar and Conversion to Buddhism.
S. W. Bakhle: Relation of Body-Mind Statements.

Subscription: Rs. 15/–; £1.50; $5.00 for Institutions.
Rs. 12/– for individuals.
Rs. 6/– for students.
Rs. 4/– for single copy.

All correspondence should be sent to:
Indian Philosophical Quarterly, Department of Philosophy,
University of Poona, POONA—7.

The Monist
An International Quarterly Journal of General Philosophical Inquiry

Founded 1888 by **Edward C. Hegeler**   Editor, Eugene FREEMAN

Editorial Board: William P. Alston, Monroe C. Beardsley, Lewis White Beck, William A. Earle, Dagfinn Føllesdal, William Frankena, Maurice Mandelbaum, R. Barcan Marcus, Richard Martin, Mary Mothersill, Joseph Owens, Richard Rorty, J. B. Schneewind, Wilfrid Sellars, John E. Smith, Richard Wasserstrom.

Managing Editor, Ann FREEMAN

EACH ISSUE IS LIMITED TO ARTICLES ON A SINGLE GENERAL TOPIC. *Communicate with the Editor in advance for Special Instructions. Papers should be 6000-8000 words in length and must be submitted in duplicate nine months prior to the scheduled publication of the issue, accompanied by return postage.*

GENERAL TOPICS for recent and forthcoming issues:

SCHEDULED PUBLICATION DATES:

Vol. 58,	No. 4	Oct., 1974	The Philosophy of Moral Education
Vol. 59,	No. 1	Jan., 1975	The Philosophy of Husserl
Vol. 59,	No. 2	Apr., 1975	Philosophical Problems of Death
Vol. 59,	No. 3	July, 1975	Language, Thought, and Reality
Vol. 59,	No. 4	Oct., 1975	The Philosophy of Mysticism
Vol. 60,	No. 1	Jan., 1976	Bioethics and Social Responsibility
Vol. 60,	No. 2	Apr., 1976	Philosophy and Religion in the Nineteenth Century
Vol. 60,	No. 3	July, 1976	New Directions in Semantics
Vol. 60,	No. 4	Oct., 1976	Historicism and Epistemology

Editorial Office: Box 1908, Los Gatos, California 95030
Business Office: Box 599, La Salle, Illinois 61301

SUBSCRIPTION RATES: United States: Annual (4 issues): Institutions, $8.00; individuals, $6.00. Singles copies: $2.25. Foreign postage: Add 15 cents to single copy rate or 60 cents to subscription rate.

DIRECTORIES OF PHILOSOPHY

The world's most comprehensive source of information on philosophy . . .

DIRECTORY OF AMERICAN PHILOSOPHERS, 1974-75

Seventh Edition, Edited by Archie J. Bahm

Information on the United States and Canada: 9300 Philosophers with Addresses, Fellowships and Assistantships, 2900 Colleges and Universities, 98 Philosophy Journals, 55 Philosophical Societies, 125 Publishers on Philosophy Books, Statistics of the Profession, Complete indices, 662 pages (Numbers based on 1972-73 Edition).

INTERNATIONAL DIRECTORY OF PHILOSOPHY, 1974-75

Third Edition, Edited by Ramona Cormier, Paul Kurtz
Richard H. Lineback and Gilbert Varet.

Information on 91 countries, excluding the U.S. and Canada: 5000 Philosophers with Addresses, 700 Colleges and Universities, 240 Institutes of Philosophy, 175 Philosophical Societies, 470 Philosophy Journals, 690 Publishers of Philosophy Books, Complete indices, 451 pages (Numbers based on 1972-73 Edition).

Publication Date: January 1974
Price of the two volume set: $35. (Individual Volumes: $20).
Philosophy Documentation Center, Bowling Green State University,
Bowling Green, Ohio 43403, U.S.A.

AMERICAN PHILOSOPHICAL QUARTERLY

Edited by
NICHOLAS RESCHER

VOLUME 11/NUMBER 2　　　　　　　　　　　　　　　　　　APRIL 1974

CONTENTS

I. KEITH LEHRER: *Truth, Evidence and Inference*

II. J. KELLENBERGER: *God and Mystery*

III. DAN CRAWFORD: *Bergmann on Perceiving, Sensing and Knowing*

IV. ERNEST SOSA: *The Concept of Knowledge*

V. WARREN S. QUINN: *Theories of Intrinsic Values*

VI. GARETH MATTHEWS: *Paradoxical Statements*

VII. CHARLES RIPLEY: *A Theory of Volition*

Annual Subscription: $18.00 to Institutions (excluding Annual Monograph, price $7.00)
$ 8.00 to Individuals (including Annual Monograph)

**Published by BASIL BLACKWELL. Orders to: 108 Cowley Road,
Oxford OX4 1JF, England.**

STUDIES IN PHENOMENOLOGY AND EXISTENTIAL PHILOSOPHY

from *Northwestern University Press*
1735 Benson Avenue
Evanston, Illinois 60201

THE WRITINGS OF JEAN-PAUL SARTRE
EDITED BY MICHEL CONTAT AND MICHEL RYBALKA
translated by Richard C. McCleary two volume set: $35.00

Volume I: A Bibliographical Life $30.00

Not just a bibliography but a biography of Sartre as well, this work lists every one of his writings, from the most ephemeral interview to the most powerful masterpiece. The annotations are incisive and frequently extensive; many entries are accompanied by excerpts both from the work cited and from the responses it evoked upon publication. Some brief works are reproduced in full. "Almost every reference is perceptively and economically situated not only in Sartre's own development but in the context of contemporary intellectual and political history. . . . This dynamic and readable bibliography . . . must be unique in its genre" (Rhiannon Goldthorpe, Books Abroad, *in a review of the French edition*). Since its French publication in 1970, the work has been corrected and updated to 1972 by Michel Rybalka, and references to English translations have been added.

Volume II: Selected Prose $10.00

This volume contains thirty-two short pieces by Jean-Paul Sartre, most of them translated here for the first time, and many of them previously inaccessible in any language. The book has a wide range, both in time and in subject matter. It begins with two satirical short stories written by the eighteen-year-old Sartre in 1923 and ends with a paper on ethics presented in 1964. Included is Sartre's first play, *Bariona*, a mystery play about the birth of Christ, written in a prisoner-of-war camp and first performed by Sartre and his fellow captives on Christmas Day, 1940, in Stalag XII D at Trèves. Among the other pieces are passages from the diary he kept as a soldier during the retreat of the French army in 1940; an angry denunciation of the American people's acquiescence in the execution of the Rosenbergs; an interview about the Algerian situation granted to Francis Jeanson in 1959; a witty riposte to an attack on existentialism in a 1947 edition of *Pravda*; and book, film, and theater reviews, philosophical essays, and commentaries on life in Europe and America.

SANSKRIT texts, translations, studies

FROM ADYAR LIBRARY & RESEARCH CENTRE, MADRAS

AVAILABLE IN THE AMERICAS FROM

The Theosophical Publishing House
AT WHEATON, ILLINOIS 60187

SRI PANCARATRA-RAKSHA OF VEDANTA DESIKA. *Pandit M. Duraiswami Aiyangar and Pandit T. Venugapalacharya (Eds.).*
A 14th century treatise on the secret worship as elaborated in the Sakta Agamas. Text in Sanskrit. Introduction in English. "Among the great scriptures of the Hindus, the Agama (also known as the Tantra) and the Nigama (well known as the Veda) are considered to be the most sacred and authoritative. ALS 36. Cat. #7482. $6.00

INDIAN CAMERALISM. *Professor K. V. Rangaswani Aiyangar.*
A comparative study of Indian and Western political theories. In English. 184 pages. Cameralism represents the ideas of a school of social thought, which had a great vogue on the continent of Europe from the 15th to the 18th centuries to which German economics is indebted. The name is derived from German *Kammer* (chamber), which was the designation of the administrative and advisory body of experts found in most German states at that time. The "Kammer" dealt not only with purely economic questions like agriculture and industry, finance, and taxation but also with police and law. The author has given the title INDIAN CAMERALISM to this study largely based on *Arthasastra*. ALS 66. Cat. #7267. $4.00

DESCRIPTIVE CATALOGUE OF SANSKRIT MANUSCRIPTS, VOLUME VIII.
This volume covers the fields of Samkhya, Yoga, Vaisesika and Nyaya. The pattern adopted for cataloguing is that prescribed by the Ministry of Education, Government of India. Tabular columns, relevant extracts, colophons in the case of rare and important works. ALS 100. Cat. #7471. $20.00

SOME CONCEPTS OF ALAMKARA-SASTRA. *Dr. V. Raghavan.*
Critical and detailed historical survey, in English, of some of the important concepts in Sanskrit Poetics. Revised edition, 1972. Chapter synopsis and index. 343 pages. ALS 33. $8.00

SRIHARICARITA MAHAKAVYA OF SRIHARI PADMANABHASASTRIN. *Ed. by Prof. T. Venkatacharyya.*
A poetical work dealing with the epic story of Sri Rama including the Uttarakanda in the Arya metre. Sanskrit text, English introduction. ALS 102. Cat. #7322. $11.50

EPISTEMOLOGY OF DVAITA VEDANTA. *Dr. P. Nagaraja Rao.*
Study of the Dvaita philosophy of Madhvacharya. A pluralistic, theistic, and realistic system based on the Puranas and the Mahabarata. Paperbound. In English. ALS 22, Parts 3-4. 120 pages. Cat. #7442. $6.00

THE PUBLISHERS OF QUEST BOOKS
The Theosophical Publishing House (U.S.A.), Box 270, Wheaton, Illinois 60187
■ 25 per cent discount to scholars.

MAIL BOOK-LENDING SERVICE
ORIENTAL SECTION — NATIONAL LIBRARY
THE THEOSOPHICAL SOCIETY IN AMERICA

■ SPECIALISTS are invited to use the Library's service for works in Asian studies, comparative religion, and philosophies. *Reading and acquisition lists (50¢) on request.* Fee $8 a year, postage additional. Address: Librarian, The Theosophical Society in America, Box 270, Wheaton, Illinois 60187. *Books loaned only in the U.S.A.*

Raïssa's Journal

presented by Jacques Maritain

Shortly after Raïssa Maritain's death from cerebral thrombosis in 1960, her grief-stricken husband, Jacques Maritain, began to go through her papers. Among them he found a set of journals Raïssa had kept over the 54 years of their life together. Jacques knew that she kept diaries, just as he did, but he had no idea of their true nature until he read them after her death.

As few people have written more starkly and vividly of the spiritual life, the initial reaction of such friends as Thomas Merton was that these notes were too powerful and too searing to be made public, even though he felt enormously privileged to read them himself. Others expressed much the same reaction. In almost every case, however, such friends had second thoughts and encouraged Jacques to allow a regular edition of *Raïssa's Journal*.

They appear here for the first time in English, in a translation corrected, revised and approved by Jacques Maritain before his death in 1973. 426 pages, six pages of photographs, and enlarged with new matter for this translation.

Jacques Maritain:
Homage in Words and Pictures

John Howard Griffin and Yves R. Simon

A beautifully produced testimonial by two of his close friends. John Howard Griffin has contributed the notes from his diary of meetings and friendship with Maritain and over thirty striking photographs, from Princeton to Toulouse, fill many a full page. Yves R. Simon shows Maritain's importance in a perceptive biography and personal history. A unique and moving momento. Size 8½" × 11"; ninety-six pages.

These two volumes will be available to friends of The Review of Metaphysics for the combined price of twenty dollars—a notable saving over the retail price. Send your reservations early. Reservations must be received before the 22nd of May.

MAGI BOOKS, INC.
33 BUCKINGHAM DRIVE – ALBANY, N. Y. 12208

IMPORTANT WORKS IN PHILOSOPHY

EXISTENTIALISM:
With or Without God by Francis J. Lescoe, S.T.L.

This book is intended as a primer for a first course in existentialism. It explores the thoughts of six of the more important existential thinkers including: Kierkegaard, Marcel, Buber, Heidegger, Sartre, and Camus. $9.50, cloth—April

CONSCIENCE:
It's Analysis and Authority

Ed. by John Donnelly, Ph.D. and Leonard Lyons, Ph.D.

This work comprises a thorough philosophical analysis of conscience and its role in forming moral decisions including the vexing question of the infallible authority of conscience. $4.95, paper

RELIGION IN CONTEMPORARY THOUGHT

Ed. by George F. McLean, OMI

Louis Dupré, Daniel Callahan, Michael Novak, Thomas J. J. Altizer and other experts delineate the origin and current impact of a wide variety of beliefs and attitudes such as atheism, secularism, humanism, relativism, and others. $4.95, paper

TRACES OF GOD IN A SECULAR CULTURE

Ed. by George F. McLean, OMI

Langdon Gilkey, Eugene Fontinell, Charles Hartshorne and other experts consider the concept which modern man has of himself and thoroughly evaluate any new signs of the transcendent—of the divine—which such a concept might imply. $5.95, paper

THE BETRAYAL OF WISDOM

by Robert J. Kreyche

"The classical and generic definition of philosophy as 'the science of all things viewed under the light of reason' is revived and vindicated. He is concerned, not with cavils over matters of tertiary importance, but with providing a synthetic and workable overview of the human situation as it exists today."
Kirkus Reviews $3.95, paper

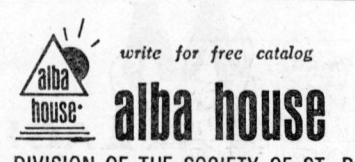

write for free catalog

alba house

DIVISION OF THE SOCIETY OF ST. PAUL
2187 VICTORY BLVD.
STATEN ISLAND, N.Y. 10314

ANALYTICAL PHILOSOPHY OF ACTION

ARTHUR C. DANTO

A study of the philosophical problems associated with the concept of basic actions. As in his earlier books, Professor Danto places the discussion in a broad historical and philosophical perspective.

From a review of his *Analytical Philosophy of Knowledge:* "An important book, written with a conscious effort to unite intellect and sense. Danto is one of the finest stylists in the business today." —*The Review of Metaphysics* $12.50

THE PROBLEM OF METAPHYSICS

D. M. MACKINNON

In this study of the nature and possibility of a revisionary metaphysics, Professor MacKinnon presents an individual and deeply considered examination of some of the most intractable problems faced by philosophers and theologians. He discusses the contributions made by Kant, Aristotle and Plato. Much of the study is devoted to the role of dialogue, myth, metaphor and parable in aesthetic and philosophical explanations. $9.50

PROBLEMS AND THEORIES OF PHILOSOPHY

KASIMIERZ AJDUKIEWICZ
Translated by Anthony Quinton and Henryk Skolimowski

An English translation of an introduction to philosophy by the late Kasimierz Ajdukiewicz which reflects the clarity, power and innovation of modern Polish philosophy. The study conveys concisely and so far as possible systematically the main problems in epistemology and metaphysics and the most historically important of the solutions proposed to them. $9.50

Cambridge University Press
32 East 57th Street, New York, N.Y. 10022

In philosophy our name's not worth dropping

(but our authors are worth picking up on)

Logic: A Philosophical Introduction
by Jack Kaminsky, *State University of New York at Binghamton,* and Alice R. Kaminsky, *SUNY College at Cortland*

Logic isn't just for math fiends. Everyone uses logic in one way or another. This book is written for people who want to know more about what's going on in logic, without having to wade through a lot of unnecessary math. It is a *philosophical* introduction to logic with the emphasis on philosophical issues of logic, rather than mathematical problems.

The authors use a very readable style throughout, and maintain a leisurely, though by no means lazy, pace. They show logic not only in relation to mathematics, but in relation to language, philosophy, literature, and even simple common sense. They're not afraid to be controversial or speculative either. The result is a book on logic that is truly for our times, one that continually stresses the connection between reasoning and the use of logic. (1974)

A Profile of Mathematical Logic
by Howard DeLong, *Trinity College, Hartford, Connecticut*

This book will appeal to anyone who wants a relatively brief and readable introduction to mathematical logic: its historical background, its nature, and its philosophical implications. The emphasis throughout is on understanding, not technique. The focus is on topics that can't be dealt with in a mechanistic fashion. These topics include: the historical reasons why Aristotelian logic came into being; how it came about that after more than two thousand years traditional logic gave way to mathematical logic; the nature of the formal axiomatic method and the reasons for its use; the main results of metatheory, and the philosophic import of these results. (304 pp, 22 illus—1970—$11.75)

College Division
ADDISON-WESLEY PUBLISHING COMPANY, INC.
Reading, Massachusetts 01867